IoT Automation

ARROWHEAD FRAMEWORK

IoT Automation

ARROWHEAD FRAMEWORK

EDITED BY

Jerker Delsing

CRC Press
Taylor & Francis Group
Boca Raton   London   New York

CRC Press is an imprint of the
Taylor & Francis Group, an **informa** business

CRC Press
Taylor & Francis Group
6000 Broken Sound Parkway NW, Suite 300
Boca Raton, FL 33487-2742

© 2017 by Taylor & Francis Group, LLC
CRC Press is an imprint of Taylor & Francis Group, an Informa business

No claim to original U.S. Government works

Printed on acid-free paper
Version Date: 20170113

International Standard Book Number-13: 978-1-4987-5675-4 (Hardback)

Visit the Taylor & Francis Web site at
http://www.taylorandfrancis.com

and the CRC Press Web site at
http://www.crcpress.com

To all colleagues in the Arrowhead project.

Contents

4   Arrowhead Framework core systems and services   89

Jerker Delsing, Jens Eliasson, Michele Albano, Pal Varga, Luis Ferreira, Hasan Derhamy, Csaba Hegedűs, Pablo Puñal Pereira, and Oscar Carlsson

5 Application system and services: Design and implementation - A cookbook   139

*Jerker Delsing, Michele Albano, Luis Ferreira, Fredrik Blomstedt, Per
Olofsson, Pal Varga, Federico Montori, and Fabio Viola*

6 Engineering of IoT automation systems   161

*Oscar Carlsson, Daniel Vera, Eduardo Arceredillo, Markus G. Tauber,
Bilal Ahmad, Christoph Schmittner, Sandor Plosz, Thomas
Ruprechter, Andreas Aldrian, and Jerker Delsing*

Erkki Jantunen, Mika Karaila, David Hästbacka, Antti Koistinen,
Laurentiu Barna, Esko Juuso, Pablo Puñal Pereira, Stéphane Besseau,
and Julien Hoepffner

9 Application system design: Complex systems management and automation   281

Michele Ornato, Tullio Salmon Cinotti, Alberto Borghetti, Paolo Azzoni, Alfredo D'Elia, Fabio Viola, Federico Montori, and Riccardo Venanzi

10 Application system design - High security   317

Andreas Aldrian, Peter Priller, Christoph Schmittner, Sandor Plosz,
Markus Tauber, Christina Wagner, Daniel Hein, Thomas Ebner,
Martin Maritsch, Thomas Ruprechter, and Christian Lesjak

11 Application system design - Smart production   331

Daniel Vera, Robert Harrison, Bilal Hamed, Claude le Pape, Chloe
Desdouits, and Hasan Derhamy

Foreword

In 1990s, I was working with Kawasaki Electric Co. in Tokyo, Japan. Among many products, the company was making high-voltage switchgear. Actually quite a few types. Some of them were sold in Europe under ASEA marketing name. Although the company's manufacturing plants had substantial capacity sufficient to fulfill long term orders, the management was facing occasionally production and scheduling dilemmas as a result of unanticipated orders to be fulfilled in a relatively short period of time. Investing in an additional factory was financially risky. The 1990s were the years of deflation in Japan. Getting an investment loan was not easy. At the same time, losing customers, especially those from abroad, was not on the cards. Switching between types of production items necessitated major adjustments on the factory floor, re-tooling, etc. Additional problem was with the supply chain providers, and availability of components needed. "Just-in time" was the strategy of the day. Going for suppliers from outside the kiretsu structure was out of question, and unheard in Japan. Being in charge of the factory automation R&D group, I was given a task to study the problem and find a comprehensive solution. The objectives were many: increased automation level to minimize time needed for factory floor changes and re-tooling, improving communication on the factory floor, better management of the component stock reserves (we developed here a predictive model reflecting global demand for switchgear equipment, and likely behavior of our customers), etc.

Specification of solutions needed came relatively easy. Though, it took much more time then we anticipated. At that time, we had no tools to help extract requirements. Interviews with key decision makers and factory operators were frequently bringing us conflicting requirements. The challenge at that stage, however was to capture the requirements specification, which was in an informal form, in a design model, or rather models, as we experimented with a few options. The final choice was stochastic Color Petri Nets, which allowed for formal verification of the design model, preliminary performance evaluation, and subsequent changes to the requirements. The next step was to partition the model into hardware and software, and adopting underpinning technologies. The underpinning technologies . . . This is were the project came to the halt. There were not many. We even tried to remedy the communication infrastructure across the enterprise. It was obvious that to integrate the shop floor level with the office level, and have seamless connectivity across plants, including suppliers, Ethernet had to be adopted and tweaked to allow for real time operation. We even came with prototypes of controllers to make Ethernet real-time. Quite a few years before the real-time Ethernet standardization process took off. But manufacturers are not in the business of developing technologies. The ambitious and elusive project had to be put to rest.

In the past twenty years, or so, since the mentioned project, ICT technologies advanced considerably bringing the vision of integration of various levels of the industrial enterprises to reality. Industrial Ethernet with its various real-time implementations can be used in the automation domain and for vertical integration of industrial enterprises at a global scale. Wireless industrial (sensor) networks with WirelessHART, ISA100.11a, and IEC62601 (WIA-PA) are slowly finding their way on the factory/plant floors all over the world - used primarily for monitoring. 6LoWPAN (IPv6 over Low-Power Wireless Personal Area Networks) allows IPv6-based communication over IEEE.802.15.4 - WirelessHART, ISA100.11a, and WIA-PA all use IEEE.802.15.4 at the physical layer level - thus allowing to link embedded nodes (controllers in field devices) to the IP based communication infrastructure. Perhaps one of the most important developments to provide an integration framework for large heterogeneous systems was that of Web Services, and particularly service-oriented architecture approach (SOA), also offering a range of solutions and protocols for the embedded device environment. CoAP (Constrained Application Protocol), an application layer protocol compatible with IPv4 networks, allows to control and integrate resource-constrained devices using embedded web services. CoAP, as 6LoWPAN, allows for networks of resource constrained devices to be integrated with the IP based communication infrastructure. All those advances, and many others not mentioned, offer a rich plethora of tools to realize integration and automation of large scale heterogeneous engineering systems such as global industrial enterprises, and automation of diverse engineering domains in general. Internet of Things is becoming a reality.

The book: "IoT Automation - Arrowhead Framework", edited by Jerker Delsing of Luleå University, is the best ever written account of a scientific and engineering effort to understand impact of the IoT concept on automation in general, irrespective of the application domain: implications, and an assessment of the existing solutions and technologies. The effort is clearly not finished. There are areas to be still further explored. It is hoped the second stage of the Arrowhead project is going to give us more answers, and particularly going to identify issues in need of solutions. The book is a must reading for anybody involved in the general automation. Particularly for those who are able to turn ideas into technology and products.

Richard Zurawski San Francisco, August 2016

Preface

The core content of this book described the outcome of the Arrowhead project. The ambition was to support the creation of large and scalable cloud based automation system. The key enabler was considered to be interoperability and integrability in-between almost any IoT devices.

After more than 3 years of hard work I think that we have achieved something which is more than that. Interoperability is achieved at a service level. Integrability is achieved through the concept of local automation clouds. Further important requirements of automation system like real time performance, security, ease of engineering and scalability is addressed. Real time and security is supported through the idea of limited and manageable local cloud. Scalability os supported by inter-cloud communication. Interestingly we have industrial partner who claim saving on application engineering time in the order of 80% or five times.

All in all I and many of my co-workers in the Arrowhead project have spent un-countable hours on the Arrowhead Framework technology, its implementation and application to a wide range of industrial automation use cases.

It my sincere hope that other with the help of this book and on-line resources will benefit from the Arrowhead Framework and possibly can build an open community around the technology such that the automation industry can take a step into the future of affordable agile and massive automation systems.

I also like to express my sincere thanks to ECSEL-JU, the European commission and the public authorities of 15 European countries and their joint financial support for the Arrowhead project. Finally I think its important to express the appreciation to the support given by both Artemis-ia and ProcessIT.EU.

August 2016

Jerker Delsing

List of Figures

List of Tables

Contributors

Jerker Delsing
Luleå University of Technology
Luleå, Sweden

Pal Varga
AITIA
Budapest, Hungary

Jens Eliasson
Luleå University of Technology
Luleå, Sweden

Fredrik Blomstedt
BnearIT AB
Luleå, Sweden

Per Olofsson
BnearIT AB
Luleå, Sweden

Hasan Derhamy
Luleå University of Technology
Luleå, Sweden

Oscar Carlsson
Midroc Automation AB
Stockholm, Sweden

Pablo Puñal Pereira
Luleå University of Technology
Luleå, Sweden

Csaba Hegedűs
AITIA Inc,
Budapest, Hungary

Arne Skou
Aalborg University
Aalborg, Denmark

Luis Ferreira
ISEP, Polytechnic Institute of Porto
Porto, Portugal

Martinez De Soria Sanchez
Tecnalia
San Sebastian, Spain

Michele Albano
ISEP, Polytechnic Institute of Porto
Porto, Portugal

Daniel Vera
Fully Distributed Systems Ltd.
Coventry, UK

Eduardo Arceredillo
Fundacion Tekniker
San Sebastian, Spain

Markus G. Tauber
Fachhochschule Burgenland GmbH
Vienna, Austria

Bilal Ahmad
University of Warwick
Coventry, UK

Christoph Schmittner
Austrian Institute of Technology
Vienna, Austria

Sandor Plosz
Austrian Institute of Technology
Budapest, Hungary

Thomas Ruprechter
Infineon Technologies AG
Graz, Austria

Andreas Aldrian
AVL List GmbH
Graz, Austria

Peter Priller
AVL List GmbH
Graz, Austria

Christina Wagner
Austrian Institute of Technology
Vienna, Austria

Daniel Hein
IAIK TU-Graz
Graz, Austria

Thomas Ebner
Evolaris Next Level
Vienna, Austria

Martin Maritsch
Evolaris Next Level
Vienna, Austria

Christian Lesjak
Infineon Technologies AG
Graz, Austria

Rodrigo Castiñeira
Indra
Barcelona, Spain

Chloé Desdouits
Schneider Electric
Grenoble, France

Thibaut Le Guilly
Aalborg University
Aalborg, Denmark

Inge Isasa
Orona
San-Sebastian, Spain

Jani Jokinen
Tampere University of Technology
Tampere, Finland

Kaspars Kondratjevs
Riga Technical University
Riga, Latvia

Nadezhda Kunicina
Riga Technical University
Riga, Latvia

Lorenzo Manero
IK4-IKERLAN
Arrasate-Mondragón, Spain

Aitor Milo
IK4-IKERLAN
San-Sebastian, Spain

Javier Monge
Indra
Barcelona, Spain

Claude Le Pape
Schneider Electric
Grenoble, France

Per D Pedersen
Neogrid
Aalborg, Denmark

Torben Bach Pedersen
Aalborg University
Aalborg, Denmark

Petur Olsen
Aalborg University
Aalborg, Denmark

Laurynas Šikšnys
Aalborg University
Aalborg, Denmark

Radislav Smid
Czech Technical University
Prague, Czech Republic

Rafael Socorro
Acciona
Madrid, Spain

Petr Stluka
Honeywell
Prague, Czech Republic

Anatolijs Zabasta
Riga Technical University
Riga, Latvia

Erkki Jantunen
VTT
Helsinki, Finland

Mika Karaila
Valmet
Tampere, Finland

David Hästbacka
Tampere University of Technology
Tampere, Finland

Antti Koistinen
Oulu University
Oulu, Finland

Laurentiu Barna
Wapice
Tampere, Finland

Esko Juuso
Oulu University
Oulu, Finland

Stéphane Besseau
Airbus
Toulouse, France

Julien Hoepffner
Airbus
Toulouse, France

Michele Ornato
CRF
Turin, Italy

Tullio Salmon Cinotti
University of Bologna
Bologna, Italy

Alberto Borghetti
University of Bologna
Bologna, Italy

Paolo Azzoni
Eurotech
Bologna, Italy

Alfredo D'Elia
University of Bologna
Bologna, Italy

Fabio Viola
University of Bologna
Bologna, Italy

Federico Montori
University of Bologna
Bologna, Italy

Riccardo Venanzi
University of Bologna
Bologna, Italy

Robert Harrison
Warwick University
Coventry, UK

1

Towards industrial and societal automation and digitisation

Jerker Delsing

Luleå University of Technology

CONTENTS

1.1 The need for new technology

High level topics in today's society are sustainability, flexibility, efficiency, and competitiveness. These in turn are driven by big societal questions like environmental sustainability, availability of energy and other raw materials, and rapidly changing market trends. We find several trends that in different ways address these topics. One is the move from large monolithic organisations towards multi-stakeholder cooperations where cooperation is fostered by market requirements. Another is learning from previous products, other parts of the value chain, the life cycle of the product and the product or service production process itself.

These trends are putting new requirements on the technology used to support product and service production in todays society. This is why new approaches to production automation and stakeholder cooperation are sought by many players. It's from these questions and requirements that the quest for digitisation of production is arising. Digesting this reveals a number of gaps regarding technology, organisation, cooperation structure, operational management, and related business models that have to be addressed.

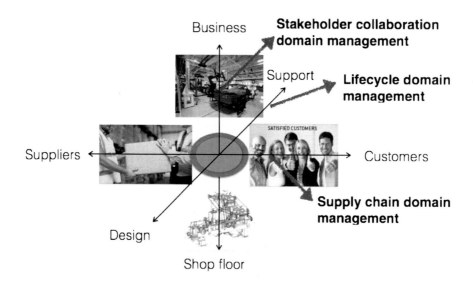

FIGURE 1.1
Three important axes for collaborative production: product life cycle, supply chain, and stakeholder integration management.

Around organisation, cooperation structure, and operational management high profile key aspects are related to three domains (see Figure 1.1):

- Product life cycle management

- Supply chain management

- Stakeholder integration management

With the move from large monolithic enterprises towards multi-stakeholder cooperation, management is changing towards distributed multi-stakeholder collaboration with distributed responsibilities and decision making. The flexible collaboration along each of the three domains also opens the possibility of dynamic learning within each of the three domains (see Figure 1.2). A further aspect is that these domains tend to become wider (longer), involving more stakeholders with diverse objectives and more details and variations of the service or product to meet customer diversity and service and product quality.

These ideas are currently emerging but are regarded as very important to address the high level topics of flexibility, efficiency, and competitiveness and with suitable incentives also supporting sustainability.

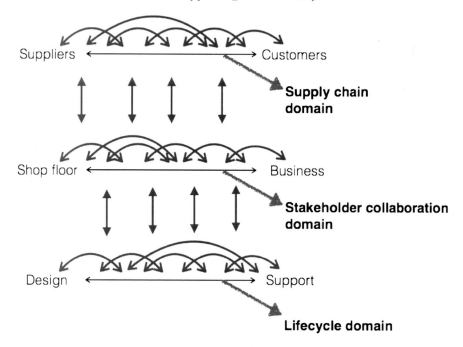

FIGURE 1.2
Learning within each of the three collaborative production domains supported by feedback and feedforward of information along the domain and between involved stakeholders.

To support these developments there are a number of technology gaps which seemingly cannot be addressed by the current state of the art. For this purpose, a number of new technologies are emerging to fill these gaps. Some current big buzz technologies are

- Internet of Things, IoT,

- System of Systems, SoS,

- Cyber Physical Systems, CPS

- Cloud

- Big data

- Service-Oriented Architecture, SOA

Despite all the new operational and organisational ideas and emerging technologies, the automation fundamentals captured by today's state of the art automation technology have to be maintained. Thus the next generation of automation and digitalisation technology has to meet a large set of requirements while also involving a wider scope of actors and stakeholders. This is the big challenge for the automation and digitisation technology suppliers of the future.

Within the Arrowhead project [1] technologies addressing the capability of building large System of Systems based on IoT using an service-oriented approach has been in focus. The results have been tested in more than 20 real-world automation and digitisation use cases, resulting in a technology framework supporting some of the most prominent requirements to automation systems.

1.2   From DCS and SCADA to Internet of Things and System of Systems

Modern industrial production and manufacturing systems have evolved in basically four generations. The first generation that enabled the industrial revolution dates back to around 1850 or so. The use of steam-powered machines enabled mass production of goods such as clothes, cars, and many other products in the beginning of the 20th century.

In the second generation, the use of efficient pneumatic systems became a widely adopted solution for mass production. The combined use of pneumatic valves and sensors enabled automatic production systems to be used in industrial applications.

The third-generation systems evolved from pneumatic to electrical motors. The use of electricity as the energy source enabled even newer types of automatic control systems to be developed. Sensors and actuators were now connected to new types of monitoring and control systems like Distributed Control Systems (DCS) and Supervisory Control and Data Acquisition (SCADA) using technologies such as field buses. The hierarchical approach of device-level, DCS, and SCADA (known a ISA-95), soon became the defacto architectural style for how industrial productions systems were designed and deployed. DCS and SCADA systems soon became networked, which enabled tight integration between control systems and Enterprise Resource Planning Systems (ERP) and Manufacturing Execution System (MES). Today this is the most widely used approach by the industry, and has been so for at least the last 20-30 years. In the 1990s the current state of the art architecture ISA-95 was established [2]. Seemingly the size of ISA-95 based automation systems seems to be limited to approximately 100.000 I/O points. This becomes a technology bottleneck in the view of the upcoming smart cities and smart energy grids.

In 2011, the concept of Industry 4.0 [3] was born in Germany. This concept builds upon the last generation of industrial monitoring and control systems, but enables an even finer level of interaction between shop-floor devices and high-level enterprise systems. In Industry 4.0, state of the art technologies like the Internet of Things (IoT) and the Cyber-Physical Systems (CPS) are utilised in order to be able to break the classical strict hierarchical approach of ISA-95 [2] with a more flexible approach without barriers and closed systems. By basing all communication on standards-based protocols, like the TCP/IP protocol suite, it is now possible to have information exchange between (almost) any systems in a production facility. This opens up the possibility for new strategies in terms of global plant optimisation, minimised energy consumption, safety and security, etc.

Of course, when more complex communications stacks are being deployed, even on resource-constrained sensors and actuators, new challenges and problems that must be handled are introduced [4]. One of the most critical challenges is security. When more and more devices are networked, it opens up vulnerabilities for remote network-based attacks. Earlier systems could only be configured by someone actually on the factory floor, but when Internet protocols are used, it can allow users to change settings and configuration from anywhere on the planet. When communication stacks became complex, the need for more complex operating systems also increased. Today, many devices are based on Linux or Windows platforms. The use of general-purpose platforms also opens up vulnerabilities for threats such as viruses and software flaws. One bug that sent shock waves throughout the computer industry was the Heartbleed bug [5], which could potentially turn any secured Linux-based system into something wide open.

The current limitations and bottlenecks are projected to be addressed by the introduction of concepts like the Internet of Things (IoT), System of Systems (SoS) and various cloud technologies.

This chapter will give an overview of automation systems and their key building blocks and associated technologies. Next is a discussion on current trends in automation systems and how IoT and SoS are expected to be used to enable the design, engineering, commissioning, and operation of very large and highly collaborative automation systems. Further will discussion how automations system can become agile to their "surroundings", allowing for local and global optimisations leading to reduced "production costs" by increasing important key performance indexes like Overall Equipment Efficiency (OEE) and raw material yield while simultaneously allowing for minimal energy and environmental footprint.

1.3   Automation system architectures

ISA 95 Levels – Distinct Sets of Activities

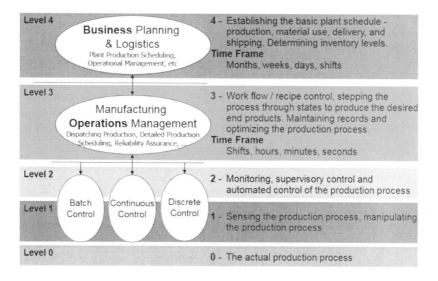

FIGURE 1.3

The ISA-95 automation pyramid architecture is the current state of the art for larger automation systems.

ISA-95 is today's standard architecture for automations systems [2]. ISA-95 is standardised through ISA [6]. Accompanying the ISA-95 standard are related standards like ISA-99 and the related IEC 62443 [7, 8] which addresses control system security.

ISA-95 can be described with the help of Figure 1.3. We find the following five layers:

- Process - level 0
  Describing the physical process to which sensors and actuators are connected and upon which closed loop or digital control is expected to act to support optimisation of the process with respect to, e.g., product quality, energy efficiency, raw material yield, OEE, etc.

- Sensor and actuators - level 1
  Sensors and actuators are connected to the physical process, level 0. They are intended to collect real-time sensor data and information about the

process and are capable of adjusting the process operation by actuation through the installed actuators.

- Monitoring, control, and supervision - level 2
  Based on sensor data/information from level 1, pursue closed loop, DCS, and digital monitoring and control, SCADA, aiming to adjust the operation of the production process through actuators.

- Manufacturing Execution System (MES), - level 3
  Work flow control based on machine, material, and other resource availability.

- Enterprise Resource Planning (ERP), - level 4
  Enterprise level planning, material supply product shipment, inventory, etc.

The sensor and actuator level has a multitude of suppliers each of them often specialised within a small number of application fields. The monitoring control and supervision level is dominated by a handful of players like ABB, Emerson, Siemens, Schneider, and Metso/Valmet. The MES level has a large number of suppliers often specialised to one or a few tasks and applications. The ERP level is dominated by one player, SAP, with a number of smaller ones like, e.g., Microsoft, IFS, . . .

1.3.1 Automation system properties

The different levels in the ISA-95 pyramid have certain system properties to support the existing control paradigm. This paradigm is based data pull. Here data is pulled from the data source at equidistant time intervals. These time intervals are determined based on system physics. The range is from sub ms to many seconds.

In current automation systems we find a range of different controller types:

- DCS - Distributed Control System

- SCADA - Supervisory Control and Data Acquisition system

- PLC - Programmable Logic Controller

Most control loops executed on these types of controllers require execution in a well-known time interval resulting in controllers pulling data on multiple I/O channels at pre-defined time intervals (cycle time) and controllers which have to perform the control computations before the next data value is requested and without hampering parallel control loops. Thus all control loops are competing for the controller resources to meet their individual cycle times. Meeting such hard real-time requirements is critical to production system performance. Deviation from the hard real-time requirements may result in degraded product quality or worst case production disaster with machine breakdown as a result.

Another property of the current automation system is that most control functionality is centralised to level 2. Further, all bindings between the controller at level 2 and sensors and actuators are determined at design time.

1.3.2   Communication within automation systems

One of the critical parts in the delivery of data to controllers at the correct time is communication latency. Thus it's interesting to have a look at the type of data/information communication technologies used within each of these levels and between these levels.

Within level 1 we do find a number of legacy communication technologies ranging from analog one way communication technologies, like a 4-20mA current loop [9], voltage level, and voltage pulses. Here the 4-20mA and voltage level communication have hard real-time performance well below milliseconds. Pulse type communication does not have real time properties due to its accumulating nature. Next are simple bus technologies like Hart [10] or ASi [11] to more advanced field bus technologies like Fieldbus or Profibus [12]. Each of these technologies has real-time capabilities with which the industry is well acquainted.

As emerging technologies, we find wireless technologies based on radio standards like, e.g., 802.15.4 [13] or 802.11 [14] and Industrial Ethernet [15]. Here the 802.15.4 based WirelessHART [16, 17] technology does exhibit real-time performance and latency guarantees can be stated. The same holds for Industrial Ethernet but not for 802.11 based communication.

All these technologies are most often used to connect to the level 2. Thus requiring level 2 devices like DCS and SCADA systems to handle a wide number of communication technologies and their specific real time properties.

Within level 2 we find communication based on traditional Internet technologies like Ethernet and 802.11 wireless communication. If real-time performance with latency guarantees is needed, Industrial Ethernet or Time Triggered Protocol technologies are used [18]. In many cases the relatively new OPC-UA standard [19] seems to have become the dominant standard for both intra level 2 and level 2 to level 3 communication.

Within level 3 we most often find Internet communication using proprietary application protocols with an emergent use of OPC-UA as a standardised communication protocol. From level 3 to level 4 we often find proprietary Internet based application protocols with an increasing use of web service technologies for integration to ERP systems.

In summary, this reveals the usage of a wide variety of communication hardware technologies and protocols combined with a nearly endless number of application protocols and standards not to mention data and information semantics and related encoding. This makes cross-layer communication nearly impossible within the current state of the art automation systems.

1.4   Current trends in automations systems

A number of clear trends related to automation systems have been stated via different road maps and initiatives like Industry 4.0 [3], the Factory of the Future road map [20] and the ProcessIT.EU road map on process automation [21]. Key aspects of these trends are discussed below.

1.4.1   Production flexibility and customisation

All of these roadmaps highlight production flexibility and customisation. They further envision the need for automation system agility where an automated production system is highly dependent on surrounding systems, like, e.g., energy supply, raw material supply, and water supply.

Another important change is the increased number of stakeholders involved in production. Here the coordination and integration between stakeholders and their respective automation systems becomes important. Stakeholders are expected to be a non-static group. Thus efficient uptake of new stakeholders or exchange of one or several stakeholders is adding complexity to what digitisation and automation platforms have to support.

There are substantial efforts put into data analytics to reveal properties of, e.g., the product, the service, the production system and enabling the closing of gaps where money is spent for no additional market benefit.

1.4.2   Very large automation systems

Another clear direction is the automation of very large systems like energy grids, thus becoming smart grids and cities are transforming to smart cities. Here the number of devices and systems involved quickly becomes very large, supporting predictions of connected devices numbering between 50 to 100'ed billion.

1.4.3   Automation system security

Last but not least, security of automation systems has become a hot topic. The Stuxnet virus [22, 23, 24] made things very clear that automation systems could be hampered with, in very sophisticated ways.

Security in IT and automation systems is a relative measure. In general, you should expect that you only can delay someone from hampering your systems automation performance. Further, it's clear that intruder interest is related to gain. Your protection is to make use of multi-layer approaches to security, not only during operation but also during design, engineering, commissioning, deployment configuration, and maintenance of your automation system.

1.4.4   Automation systems are physically local systems

In addition to these global requirements for next generation automation systems, we also highlight another observation.

For most automation systems the things and systems to be controlled are in close proximity to where important system data can be gathered. In other words, the sensor and the actuator in a control loop are in close proximity to each other. Thus the real-time requirements related to control have to be fulfilled between the point of data measurement and the point of actuation. In addition, systems, to an increasing extent, must be organised to enable both local (real-time) control and global operations management.

1.4.5   Automation engineering

A clear trend is that massive automation can be made to improve production efficiency and flexibility while reducing raw material cost, energy cost, environmental footprint, and simultaneously increasing OEE. One bottleneck to increasing the number of devices involved in automation systems is the engineering cost.

1.5   Future automation system requirements

We can summarise the above trends in automation systems in a number of high-level requirements, listed here:

- Automation systems will become very large, in numbers of integrated sensors and actuators, and will integrate across a large number of stakeholders to support digitisation. Thus system scalability becomes an important feature of future automation systems. Scalability has to be dynamic in runtime. Another requirement is evolutionary behaviour, moving from a rip and replace approach over a technology migration approach to an evolutionary approach where strong parts survive and weaker parts die. One important issue for scalability and evolutionary behaviour is technology interoperability between involved devices and stakeholders. Interoperability issues to consider concern, for example:

    - Communication hardware
    - Communication protocols
    - Information semantics
    - Data encoding
    - Data compression

- Security in automation systems has to be considered over the whole life cycle of automation systems. Such security has to address, for example:
  - Authentication
  - Authorisation
  - Encryption
  - Intrusion detection

- The engineering cost for automation systems has to be reduced by orders of magnitude. This includes engineering related to:
  - Low-level automation functionality
  - Integration of automation functionalities to systems
  - System of Systems integration

- Automation systems will be very distributed but still have to meet hard real-time constrains.

In the following the usage of IoT and SoS technologies to form automation architectures and technologies supporting global requirements will be discussed.

1.6 Next generation automation and digitisation technology — IoT and SoS

The trends and perspectives put forward indicate that the current solutions used to build automation systems are not sufficient. Further, the cost connected with the engineering and building of larger automation systems involving multiple stakeholders seems to be prohibitively high.

With automation and digitisation expected for competitively, flexibility, and sustainability new technology is needed. Here IoT and SoS and their usage in automation systems is currently attracting much attention.

For the last 10 years, discussion on the next generation SCADA, DCS, and MES systems has occurred. A multitude of research projects on the topic have been executed. Some more prominent are SOFIA, SOCRADES [25], and IMC-AESOP [26]. All of them were investigating a move from a hierarchical ISA-95 approach to a more cloud-like approach. An illustration of the move from the pyramid and it's hierarchical implementation to Industri4.0 and RAMI4.0 [27] implemented using Internet and cloud technology is depicted in Figure 1.4. This is currently a changing landscape but some other efforts touching this field are, e.g., FiWare [28]. In addition, there are a growing number of cloud

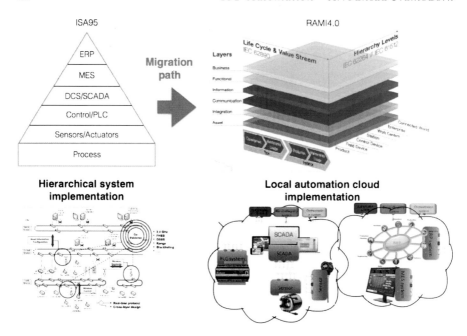

FIGURE 1.4

Transferring the ISA-95 automation pyramid architecture to the Industry4.0 and RAMI4.0 architecture. Technology for this has been investigated by several larger EU projects like, e.g., SOCRADES and IMC-AESOP.

offerings on the market. An analysis of the different approaches in 2015 can be found in [29].

In all these cases the key technology for creating integration within and between different levels of the ISA-95 architecture is Service-Oriented Architecture, SOA [30]. SOA was originally developed by IBM to enable data/information exchange between different lines of computer systems.

For the transfer of the ISA-95 architecture to a cloud-based approach, we do find some important published work regarding system architecture [31, 32], suitable technologies [33], real-time [34, 35] migration from legacy systems [36, 37], and engineering for cloud based automation [38, 39].

A parallel discussion is ongoing for the MES and ERP levels. Some important publications on cloud approaches for the MES and ERP levels are [31, 40].

For cloud-based automation, other important technology developments are the Internet of Things (IoT) and System of Systems (SoS). The properties of IoT and SoS based automation systems based on SOA are discussed in more detail below.

1.6.1 Internet of Things — IoT

In general the IoT concept enables an Internet access to any type of "device/thing". An interesting review on IoT is by Höller et.al. [41]. Thus each "thing" will have an IP address and be addressable using standard Internet technology like DNS.

In an automation context, an Internet of Thing will be a device or a functionality that we can find in standard ISA-95 automation implementations. Automation applications of the Internet of Things typically include:

- Sensors

- Actuators

- Programmable logic controllers (PLCs)

- Control loops

- Data analytics

- Control models

- Simulators

- Optimisers

Thus the thing can be viewed as a physical device or as a functionality that is computed in a software system executed on any type of device having sufficient computational resources.

Currently there is no consistent technology that can be attributed to an Internet of Thing. Several standardisation efforts are under way. For example, ITU has a standardisation working group SG20 on "IoT and its applications including smart cities and communities" [42], IEEE has an ongoing IoT Ecosystem study and the IEEE P2413 draft standard for an Architectural Framework for the Internet of Things [43]. Another initiative is pushed by the IPSO alliance [44] where Internet technology IETF standards like 6LoWPAN [45] and CoAP [46] are developed and supported.

1.6.2 System of Systems — SoS

System of Systems denotes the concept of integrating single systems into larger systems. Here we can consider an IoT as a system. Maier published in 1998 five main characteristics of the System of Systems [47]. These are useful in distinguishing very large and complex but monolithic systems from true System of Systems:

- Operational Independence of the Elements: If the System of Systems is disassembled into its component systems, the component systems must be able to usefully operate independently. The System of Systems is composed of systems which are independent and useful in their own right.

- Managerial Independence of the Elements: The component systems not only can operate independently, they do operate independently. The component systems are separately acquired and integrated but maintain a continuing operational existence independent of the System of Systems.

- Evolutionary Development: The System of Systems does not appear fully formed. Its development and existence is evolutionary, with functions and purposes added, removed, and modified with experience.

- Emergent Behaviour: The system performs functions and carries out purposes that do not reside in any component system. These behaviours are emergent properties of the entire System of Systems and cannot be localised to any component system. The principal purposes of the Systems of Systems are fulfilled by these behaviours.

- Geographic Distribution: The geographic extent of the component systems is large. Large is a nebulous and relative concept as communication capabilities increase, but at a minimum it means that the components can readily exchange only information and not substantial quantities of mass or energy.

It currently seems clear that SoS architectures and solutions will be built using IoT devices. Thus the resulting SoS solutions will have a clear dependency on the specific technologies supported by the IoT devices involved.

In an automation context, SoS architectures and solutions have to meet a number of critical automation requirements. Due to the technology dependencies of the involved IoT devices SoS solution will exhibit limitations related to the IoT supported technologies.

For the design and integration of SoS based automation systems, there are a number of fundamental approaches possible. Each of these has certain fundamental features, and pros and cons.

- SOA
  Loosely coupling, late binding, autonomy, pull and push behaviour, several standardised SOA protocols, data structures, information semantics, data encryption and data encoding technologies, bulky data messages

- Agent technology
  Late binding, autonomy, degree of standardisation of agent technologies, standardised data semantics, agents introduce an intermediate data mediator between two systems

- Middleware solutions
  Very low degree of standardisation, introduction of a intermediate layer which has to be trusted

A clear majority of current approaches to IoT and SoS based automation systems have selected SOA as the approach for IoT interoperability and SoS integration. This is also the case for the Arrowhead project and SOA will thus be the approach throughout this book.

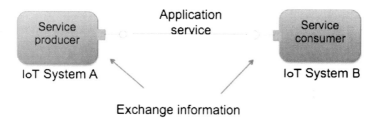

FIGURE 1.5
SOA is characterised by a service-based data exchange between a service producer system and a service consumer system.

1.6.3 Service-Oriented Architecture — SOA

SOA in general is about the data exchange between a service producer and a service consumer (see Figure 1.5). SOA exhibits a number of fundamental properties, defined in more detail below:

- Loosely coupled
  Two SOA systems do not need to know about each other at design time to allow a run-time data exchange. The identification of available system services is established at run-time making use of, service registry and discovery mechanisms supported by a service registry and discovery system. A new SOA service will register itself in the service registry upon which it will be discoverable by any other service in the network (see Figure 1.6).

- Late binding
  In a SOA system, the exchange of data between two systems is established in run-time. The run-time coupling is initiated by an orchestration mechanism possibly with the support of authentication and authorisation mechanisms supported by a orchestration system which provides the end point of the selected producer to the requesting consumer (see Figure 1.7). If necessary the authorisation system is consulted to check if the service consuming system can be authenticated and authorised to consume the requested service.

- Autonomy
  Each device and related software system can act on its own regardless of other systems. Thus it's responsible for its own data and functionalities. Once a service exchange is set up between two systems, this exchange may go on without further involvement of any supporting services/functionalities.

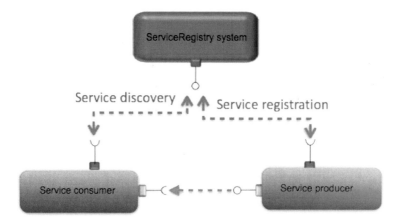

FIGURE 1.6

The loose coupling between software systems in SOA is supported by an independent service registry and discovery system, which allows systems to register services and subsequently discover the registered services.

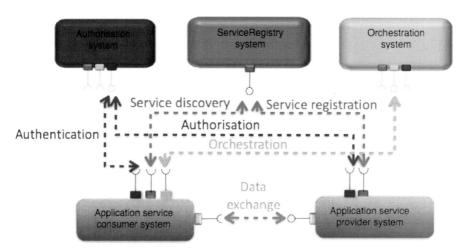

FIGURE 1.7

The late binding between software system in SOA is supported by an independent orchestration service possibly supported by a authentication and authorisation services. Thus facilitating the necessary endpoint information and authentication and authorisation allowing autonomous data exchange.

- Pull and push behaviour

  In a SOA environment, data exchange can be initiated by a service consumer requesting data — a pull behaviour. A pull behaviour can, for example, be controlled by a timer at the service consumer, thus creating data pulling of a sensor every 100 ms. The data exchange can also be initiated by a producer that knows about conditional data requests — a push behaviour. This is initiated by a data subscription under certain criteria. For example, a consumer only like pressure data whenever the pressure reading is higher than 2 bar. When this condition is meet data is pushed from the producer to the consumer.

- Existing standardised SOA protocols
  There are a number of existing SOA protocols. Each of them designed with an application in mind. A number of the most actively used SOA protocols are

  - REST [48, 49]
    Representational state transfer (REST) is the software architectural style of the World Wide Web. Thus we may regard REST more as an architecture than a protocol. It's based on HTTP and supports a few commands over HTTP enabling SOA properties.

  - OPC-UA [50]
    Supported by the OPC foundation [51]. Strong usage in automation industry. Mainly from SCADA/DCS level and upwards in the ISA-95 architecture.

  - XMPP [52]
    Supported by the XMPP community. Used by large companies for Push notifications, instant messaging and chat.

  - CoAP [53]
    Supported by the IPSO alliance formed by major players like, e.g., Ericsson, ARM, Cisco, Bosch, Intel, Oracle, HP, Google, with main application to resource constrained embedded systems — IoT devices. CoAP is REST like.

  - MQTT [54]
    MQTT is standardised by OASIS [55]. It was invented by IBM for M2M communication. The design principles are to minimise network bandwidth and device resource requirements whilst also attempting to ensure reliability and some degree of assurance of delivery.

  - DPWS [56]
    DPWS is standardised by OASIS [57]. DPWS was introduced by Microsoft as a part of their .NET suite of tools and protocols.

  - uPnP [58]
    uPnP is supported by the Open Connectivity Foundation [59] with early adopters like Philips and Samsung. Consumer product communication has been the target field of applications. We find it widely used in home appliances like TV sets and streaming devices.

- Data structures

In some protocols and standards, data structures for use with SOA protocols have been defined. One good example is the data structure defined by the OPC-UA protocol. Here 16 data models are defined mainly supporting production automation [50].

- Information semantics
  Information semantics is a vast field. There exists a wide range of semantics for almost every thing. Some examples are sensor data semantics SenML [60] and SensML [61]. Both of them address sensor data semantics but are quite different. Incompatibility between defined and in many cases standardised semantics is currently a big headache to interoperability within IoT and SoS development.

- Data encryption
  The different SOA protocols have varying levels of security support. We find a wide usage of TLS and DTLS encryption [62] by several of the SOA protocols. There are also good arguments for using protocol-independent data encryption like IPSec [63] and link-layer encryption. Another widely adopted solution is the use of Virtual Private Networks (VPN) [64].

- Data encoding technologies
  Many SOA protocols encode their data using XML [65] or JSON [66]. Both of them human readable and thus also have a tendency to become verbose. This is addressed by compression where XML compression is standardised using EXI [67, 35], with a binary representation of JSON called CBOR [68] gaining popularity. Automatic machine supported translation between JSON and XML is also available [69].

1.7   The local automation cloud approach

We currently see strong development of cloud technology and the usage of the cloud for various applications. We also see the introduction of terms like fog computing, edge computing. Here the idea is to provide local computational heads which can be used as cloud computing resources by devices connected to the head. For example, a telecom base station can be such a computational head.

To address the strong real-time, specific engineering, and security requirements of automation systems, we introduce the concept of local clouds. This enables implementations to meet strong requirements on protecting the automation functionalities from, e.g., network traffic that changes communication latency and network security attacks. It is further foreseen that integrated automation systems will comprise much larger number of actors, devices and systems than today. Thus scalability becomes important. Together this puts a strong push on engineering costs.

The local cloud concept takes the view that some specific automation actions should be supported. These actions have strong requirements on real time, ease of engineering, operation and maintenance plus security. The local cloud idea is to let the local cloud be self contained and only involve the devices and systems required to perform the desired automation tasks. Thus viewing the cloud border as a protective fence towards any other internet communication [70]

Within each of these local clouds the necessary infrastructure for building IoT based SoS automation is provided. This at the minimum includes:

- Service registry services

- Orchestration services

- Authorisation services

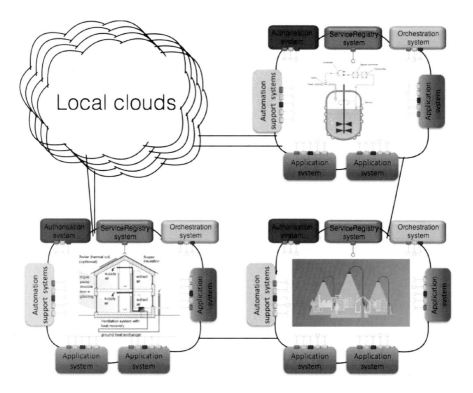

FIGURE 1.8
The local cloud concept is visualised here as a number of local clouds with largely independent automation tasks. Higher-level cloud coordination is supported by inter-cloud communication.

Each cloud can be complemented by a number of automation supportive services. Together with a number of application systems and their services the desired automation application functionality can be engineered, deployed, operated, and maintained in run-time.

I'ts obvious that many applications may require multiple local clouds to build a total automation system. Thus inter-cloud communication has to be supported while maintaining local cloud internal properties. In Fig. 1.8, the local cloud concept is depicted.

This local cloud concept will be further detailed and discussed in Chapter 2. A local cloud architecture is detailed in Chapter 3. The necessary local cloud functionalities and services to gather with automation supportive services are detailed in Chapter 4. Chapter 5 provides a cook book outlining the process of defining and implementing automation application services and how these can be deployed in a local cloud.

Bibliography

[1] "Arrowhead project." [Online]. Available: http://www.arrowhead.eu

[2] B. Scholten, *The Road to Integration: A Guide to Applying the ISA-95 Standard in Manufacturing.* ISA, 2007.

[3] J. Lee, B. Bagheri, and H.-A. Kao, "A cyber-physical systems architecture for industry 4.0-based manufacturing systems," *Manufacturing Letters*, vol. 3, pp. 18–23, 2015.

[4] J. Eliasson, J. Delsing, H. Derhamy, Z. Salcic, and K. Wang, "Towards industrial internet of things: An efficient and interoperable communication framework," in *Proceedings IEEE ICIT 2015*, March 2015, pp. 2198–2204.

[5] Z. Durumeric, J. Kasten, D. Adrian, J. A. Halderman, M. Bailey, F. Li, N. Weaver, J. Amann, J. Beekman, M. Payer *et al.*, "The matter of heartbleed," in *Proceedings of the 2014 Conference on Internet Measurement Conference.* ACM, 2014, pp. 475–488.

[6] (2016). [Online]. Available: https://www.isa.org

[7] (2016). [Online]. Available: http://isa99.isa.org/ISA99%20Wiki/Home.aspx

[8] K. S. et.al., "ISA 62443 − 4 − 2 security for industrial automation and control systems technical security requirements for iacs components," ISA, Draft Standard Draft 2 edit 4, July 2015.

[9] (2016). [Online]. Available: https://en.wikipedia.org/wiki/Current_loop

[10] R. Bowden, "Hart technical overview," HART Communication Foundation Std., Tech. Rep., 2015.

[11] (2016). [Online]. Available: https://en.wikipedia.org/wiki/AS-Interface

[12] E. Tovar and F. Vasques, "Real-time fieldbus communications using profibus networks," *IEEE Transactions on Industrial Electronics*, vol. 46, no. 6, pp. 1241–1251, 1999.

[13] "IEEE 802.15.4 standard," 2016. [Online]. Available: http://www.ieee802.org/15/pub/TG4.html

[14] (2016) IEEE 802.11 standard. [Online]. Available: http://www.ieee802.org/11/

[15] T. Skeie, S. Johannessen, and Ø. Holmeide, "Timeliness of real-time ip communication in switched industrial ethernet networks," *IEEE Transactions on Industrial Informatics*, vol. 2, no. 1, pp. 25–39, 2006.

[16] "Wirelesshart specification 75: Tdma data-link layer, rev. 1.1." HART Communication Foundation Std., Tech. Rep. hCF_SPEC-75, 2008.

[17] D. Chen, M. Nixon, and A. Mok, *Why WirelessHART.* Springer, 2010.

[18] H. Kopetz and G. Bauer, "The time-triggered architecture," *Proceedings of the IEEE*, vol. 91, no. 1, pp. 112–126, 2003.

[19] W. Mahnke, S.-H. Leitner, and M. Damm, *OPC Unified Architecture.* Springer, 2009.

[20] "Factories of the future multi-annual roadmap for the contractual ppp under horizon 2020," EFFRA, Tech. Rep., 2013.

[21] "European roadmap for industrial process automation," ProcessIT.EU, Tech. Rep., 2013.

[22] N. Falliere, L. O. Murchu, and E. Chien, "W32. stuxnet dossier," *White paper, Symantec Corp., Security Response*, vol. 5, 2011.

[23] R. Langner, "Stuxnet: Dissecting a cyberwarfare weapon," *Security & Privacy, IEEE*, vol. 9, no. 3, pp. 49–51, 2011.

[24] J. P. Farwell and R. Rohozinski, "Stuxnet and the future of cyber war," *Survival*, vol. 53, no. 1, pp. 23–40, 2011.

[25] (2016). [Online]. Available: http://www.socrades.net

[26] A. W. Colombo, T. Bangemann, S. Karnouskos, J. Delsing, P. Stluka, R. Harrison, F. Jammes, and J. L. Lastra, "Industrial cloud-based cyber-physical systems," *The IMC-AESOP Approach*, 2014.

[27] P. Adolphs and U. Epple, "Status report: Rami4.0," VDI/VDE-Gesellshact Mess- und Automatisierungstechnik, Tech. Rep., June 2015.

[28] Wikipedia, "FiWare — wikipedia, the free encyclopedia," 2016, [Online; accessed 21-April-2016]. [Online]. Available: https://en.wikipedia.org/w/index.php?title=FIWARE&oldid=711836994

[29] H. Derhamy, J. Eliasson, J. Delsing, and P. Priller, "A survey of commercial frameworks for the internet of things," in *Proceedings of Emerging Technologies & Factory Automation (ETFA), 2015.* IEEE, 2015, pp. 1–8.

[30] T. Erl, *SOA Principles of Service Design (The Prentice Hall Service-Oriented Computing Series from Thomas Erl).* Upper Saddle River, NJ, USA: Prentice Hall PTR, 2007.

[31] S. Karnouskos and A. W. Colombo, "Architecting the next generation of service-based scada/dcs system of systems," in *Proceedings IECON 2011,* Melbourne, Nov. 2011, p. 6.

[32] F. Jammes and H. Smit, "Service-oriented paradigms in industrial automation," *Industrial Informatics, IEEE Transactions on,* vol. 1, no. 1, pp. 62–70, Feb 2005.

[33] F. Jammes, B. Bony, P. Nappey, A. W. Colombo, J. Delsing, J. Eliasson, R. Kyusakov, S. Karnouskos, P. Stluka, and M. Till, "Technologies for soa-based distributed large scale process monitoring and control systems," in *IECON 2012-38th Annual Conference on IEEE Industrial Electronics Society.* IEEE, 2012, pp. 5799–5804.

[34] G. Candido, A. W. Colombo, J. Barata, and F. Jammes, "Service-oriented infrastructure to support the deployment of evolvable production systems," *IEEE Transactions on Industrial Informatics,* vol. 7, no. 4, pp. 759–767, Nov 2011.

[35] R. Kyusakov, P. P. Pereira, J. Eliasson, and J. Delsing, "Exip: a framework for embedded web development," *ACM Transactions on the Web (TWEB),* vol. 8, no. 4, p. 23, 2014.

[36] J. Delsing, J. Eliasson, R. Kyusakov, A. W. Colombo, F. Jammes, J. Nessaether, S. Karnouskos, and C. Diedrich, "A migration approach towards a soa-based next generation process control and monitoring," in *IECON 2011 - 37th Annual Conference on IEEE Industrial Electronics Society,* Nov 2011, pp. 4472–4477.

[37] J. Delsing, F. Rosenqvist, O. Carlsson, A. W. Colombo, and T. Bangemann, "Migration of industrial process control systems into service oriented architecture," in *IECON 2012.* IEEE, 2012.

[38] A. Jain, D. Vera, and R. Harrison, "Virtual commissioning of modular automation systems," in *Intelligent Manufacturing Systems*, vol. 10, no. 1. IFAC, 2010, pp. 72–77.

[39] N. Kaur, C. S. McLeod, A. Jain, R. Harrison, B. Ahmad, A. W. Colombo, and J. Delsing, "Design and simulation of a soa-based system of systems for automation in the residential sector," in *IEEE ICIT 2013*. IEEE, 2013.

[40] S. Karnouskos, A. W. Colombo, F. Jammes, J. Delsing, and T. Bangemann, "Towards an architecture for service-oriented process monitoring and control," in *IECON 2010-36th Annual Conference on IEEE Industrial Electronics Society*. IEEE, 2010, pp. 1385–1391.

[41] J. Höller, V. Tsiatsis, C. Mulligan, S. Karnouskos, S. Avesand, and D. Boyle, *From Machine-to-Machine to the Internet of Things: Introduction to a New Age of Intelligence*. Elsevier, 2014.

[42] (2016) Internet of things global standards initiative. [Online]. Available: http://www.itu.int/en/ITU-T/studygroups/2013-2016/20/Pages/default.aspx

[43] (2016) Iot related standards, projects and ecosystem study. [Online]. Available: http://standards.ieee.org/innovate/iot/

[44] (2016) IPSO alliance. [Online]. Available: http://www.ipso-alliance.org

[45] Z. Shelby and C. Bormann, *6LoWPAN: The wireless Embedded Internet*. John Wiley & Sons, 2011, vol. 43.

[46] Z. Shelby, K. Hartke, and C. Bormann, "Rfc7252 — the constrained application protocol (coap)," IETF, Tech. Rep., 2014.

[47] M. Maier, "Architecting principles for systems-of-systems." *Systems Engineering*, vol. 1, no. 4, pp. 267–284, 1998.

[48] C. Pautasso, E. Wilde, and R. Alarcon, Eds., *REST: Advanced Research Topics and Practical Applications*. Springer, 2014.

[49] (2016) Representational state transfer. [Online]. Available: https://en.wikipedia.org/wiki/Representational_state_transfer

[50] W. Mahnke, S.-H. Leitner, and M. Damm, *OPC Unified Architecture*. Springer, 2009.

[51] "OPC fondation." [Online]. Available: https://opcfoundation.org/

[52] (2016) Xmpp is the open standard for messaging and presence. [Online]. Available: http://xmpp.org

[53] Z. Shelby, "The constrained application protocol (coap) - rfc 7252," IETF, Tech. Rep., 2014.

[54] (2016) Mqtt is a machine-to-machine (m2m)/"internet of things" connectivity protocol. [Online]. Available: http://mqtt.org

[55] A. Banks and R. Gupta. (2016) Mqtt version 3.1.1. oasis standard. http://docs.oasis-open.org/mqtt/mqtt/v3.1.1/mqtt-v3.1.1.html.

[56] (2016) Devices profile for web services. [Online]. Available: https://en.wikipedia.org/wiki/Devices_Profile_for_Web_Services

[57] D. Driscoll and A. Mensch, "Oasis devices profile for web services (dpws) version 1.1," OASIS, Tech. Rep., 2009. [Online]. Available: http://docs.oasis-open.org/ws-dd/dpws/1.1/os/wsdd-dpws-1.1-spec-os.pdf

[58] (2016) Universal plug and play. [Online]. Available: https://en.wikipedia.org/wiki/Universal_Plug_and_Play

[59] (2016) Open connectivity foundation. [Online]. Available: http://openconnectivity.org

[60] C. Jennings and Z. Shelby, "Media types for sensor markup language (senml) draft-jennings-senml-10," IETF, Tech. Rep., 2013.

[61] M. Botts and A. Robin. (2014) Ogc sensorml: Model and xml encoding standard.

[62] E. Rescorla and N. Modadugu, "Datagram transport layer security rfc - rfc 4347," IETF, Tech. Rep., 2006.

[63] (2016) Internet protocol security (ipsec). [Online]. Available: https://en.wikipedia.org/wiki/IPsec

[64] Wikipedia, "Virtual private network — wikipedia, the free encyclopedia," 2016, [Online; accessed 10-April-2016]. [Online]. Available: https://en.wikipedia.org/w/index.php?title=Virtual_private_network&oldid=714464794

[65] (2016) Extensible markup language - xml. [Online]. Available: https://en.wikipedia.org/wiki/XML

[66] (2016) Introducing json. [Online]. Available: http://www.json.org

[67] D. Peintner and S. Pericas-Geertsen, "Efficient xml interchange (exi) primer," W3C, Tech. Rep., 2014.

[68] C. Bormann and P. Hoffman, "Rfc7049 — concise binary object representation (cbor)," IETF, Tech. Rep., 2013.

[69] (2016) List of converter tools json - xml. [Online]. Available: http://www.freeformatter.com/converters.html

[70] J. Delsing, J. Eliasson, J. van Deventer, H. Derhamy, and P. Varga, "Enabling iot automation using local clouds," in *Proceedings World Forum - IoT 2016*. IEEE, Dec. 2016.

2

Local automation clouds

Jerker Delsing

Luleå University of Technology

Pal Varga

AITIA Inc

CONTENTS

2.1   The local cloud concept

As discussed in the previous chapte,r we see strong development of cloud technology and the usage of cloud technology for various applications. A survey of different cloud concepts currently offered can be found in [1]. Most of these are variations of a global cloud concept where a number of requirements falls short. For digitisation and automation with their strong requirements of real-time, specific engineering, security and multi-stakeholder integration, the work of the Arrowhead project has led to the introduction of the local cloud concept. The major requirements for this are

- Automation requirement on latency guarantee for communication and control computations

- Scalability of automation systems enabling very large integrated automation systems

- Multi-stakeholder integration and operations agility

- Security and related safety of automation systems

- Ease of application engineering

The local cloud concept takes the view that specific geographically local automation tasks should be encapsulated and protected. These tasks have strong requirements on real time, ease of engineering, operation and maintenance, and system security and safety. The local cloud idea is to let the local cloud include the devices and systems required to perform the desired automation tasks, thus providing a local "room" which can be protected from outside activities [70]. In other words the cloud will provide a boundary to the open internet, thus aiming to protect the internal of the local cloud from the open internet.

This chapter will discuss which properties a local automation cloud needs to have to enable a wide variety of automation functionalities and related non-functional properties.

It's obvious that many applications may require multiple local clouds to build a large integrated automation system. Thus inter-cloud service exchanges have to be supported. Inter-cloud service exchanges often can not guarantee latency. Still the security and engineering simplicity properties of a local cloud should be preserved. In Figure 2.1 the local cloud concept is depicted.

2.2   Local cloud properties

An Arrowhead Framework local cloud should provide a number of properties important to automation. Some of these properties are related to cloud technology as such. Others are related to real time, engineering, security, scalability, and functionality. With reference to the discussion in Chapter 1, these properties are summarised as

- Self contained - now external resources needed to establish the local cloud.

  - Device, system and service registry

  - Service orchestration - SoS run-time configuration

  - Service authentication and authorisation

- Automation support

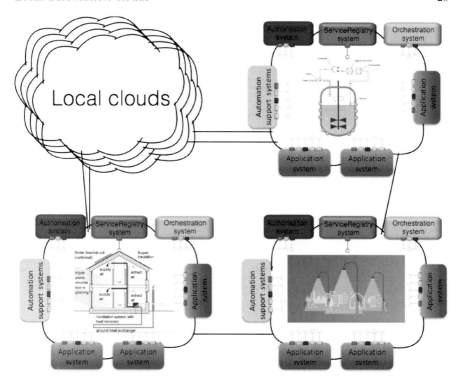

FIGURE 2.1
The local cloud concept is visualised here as a number of local clouds with
largely independent automation tasks. Higher-coordination is supported by
inter-cloud communication.

- Support for automation system design, configuration, deployment,
  operation, and maintenance
- Enabling event based information exchange
- Enabling information exchange audit
- Support for communication Quality of Service (QoS)

- Provide a security fence to external networks

- Secure bootstrapping and software updates

- Support for device, systems and service metadata

- Support for protocol and semantics transparency

- Support for secure administration and data exchange with external resources

To address the above requirements, it is here proposed, that all these properties can be addressed through the usage of service exchanges between systems using a Service-Oriented Architecture (SOA) paradigm [2]. The Arrowhead Framework aims to provide open source technology to address the above stated local properties. The Arrowhead Framework is an initiative based on the achievement of the Arrowhead project [3].

The Arrowhead Framework defines three different classes of services:

- Mandatory core services

- Automation support core services

- Application services

This chapter describes at a high level which properties a local cloud should have. In Chapter 3 a local automation cloud architecture is provided. To implement such architecture, mandatory and automation support core services are detailed in Chapter 4.

Application services are application specific. In Chapter 5 an application service design guide is providede and in Chapters 7 to 11 a number of real world application service are described. These applications are designed, deployed and evaluated in real-world automation implementations.

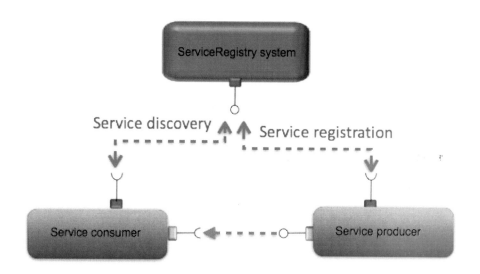

FIGURE 2.2
A service producer has to register its services with the service registry. A service consumer will then be able to discover all registered services.

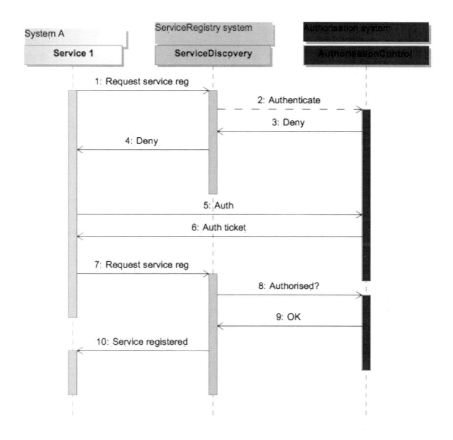

FIGURE 2.3
Registering a service to a local cloud may require the authentication of the system producing and authorising the registration of the service. The graph indicates such an authentication and authorisation process.

2.3 Local cloud establishment

In a SOA context a few cloud properties are necessary to create a service cloud. First a system that provides a service has to register the service with the cloud. Secondly, it should be possible to discover all registred services. The basics are depicted in Figure 2.2.

The registration of any service to a local cloud should be possible to restrict. Thus the service registry has to consult the authorisation services before allowing the service registration to the local cloud. This is illustrated in

Figure 2.3. This gives a way to prevent unwanted or malicious systems to register services with the local cloud.

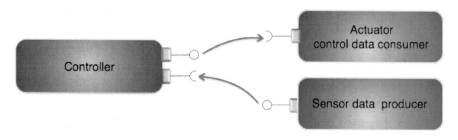

FIGURE 2.4

A control loop consisting of three systems are connected via services to create a closed control loop.

Let's assume the we register a number of services to create an automation control loop. For example, sensor data has to be provided to a controller which will output set-points to an actuator. In a local cloud this is enabled by a sensing service providing data to a controller service which in turn provides an actuation set-point service to an actuation service (cf. Figure 2.4).

Services to be included in such an control loop service exchange have to be initiated. In a local automation cloud this is handled by an orchestration service. There may be two types of orchestration information distribution. The orchestrator may push orchestration information to the systems that should consume certain services. The other way is that a system may request orchestration information. Thus it will pull this from the orchestration system (cf. Figure 2.5).

In an orchestrated control loop it's obvious that the systems involved can be identified and authenticated and that for each service exchange the service consumer can be authorised for that specific service exchange. Thus to provide internal security within a local cloud the service consumer system has to be authenticated and it's consumption of a published service has to be authorised before the service exchanges can be established. Thus authorisation services are important to provide in a local automation cloud.

The described here local cloud services, are the minimal set of services necessary to create a self contained local cloud having internal security mechanisms. In the Arrowhead Framework these services are provided by

- ServiceRegistry system

- Authorisation system

- Orchestration system

The systems and associated services are described in more detail in Sections 3.4.1 and 4.2.

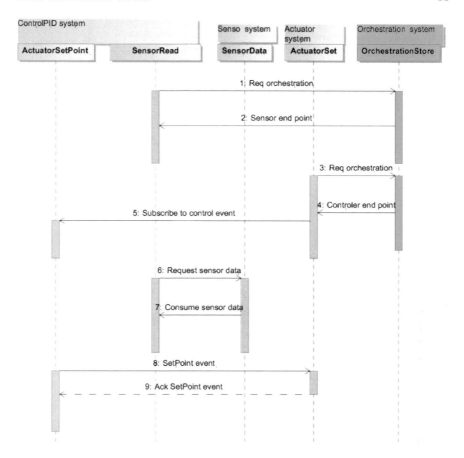

FIGURE 2.5

Control loop specific service exchanges are determined through the orchestration. Thus orchestration is an essential functionality of a local automation cloud. This sequence diagram shows a controller pull of sensor data and the actuator subscribing to a control event, a data push from the controller to the actuator if a change is necessary.

2.4 Automation support

A local automation cloud has to support smooth engineering for complex automation functionalities. Several standards and models exist addressing complex system automation. Examples are ISA95, ISA99, IEC81346, IEC61360, CAEX/IEC62424 standards and the Rami4.0 model [4, 5, 6, 7, 8, 9]. The ambition is that the properties of a local cloud should enable applications to implement solutions meeting standards like this. Thus a local cloud can have automation support addressing:

- automation system design, configuration, deployment, operation, and maintenance at
  - Plant level
  - Device level
  - Device to device interaction
  - Function level

- event based information exchange

- information exchange audit

- communication QoS

- secure inter-cloud information exchange

Each of such support systems and associated services can be deployed in a local cloud. Each system is expected to work internally in a local cloud but will be able to exchange services with systems located within the local cloud and with systems in another local cloud. Such inter-cloud information exchange has to provide secure discovery, orchestration, authorisation, and data exchange.

2.5   Automation application engineering in local clouds

The engineering of an automation system is today based on the ISA-95 automation pyramid. Most likely, for the time being this will be the basis for automation systems built using a local cloud approach.

The automation system has then to be partitioned into a number of local clouds. The criteria for such partitioning will be related to

- automation functional issues

- real-time and latency requirements

- security and safety issues

- engineering complexity related to real time and security

- deployment and test issues

Thus within a local automation cloud a certain control application will be engineered with a limited number of systems and their services involved. The systems and services will be used to build the desired control functionality. The design and implementation of the control application will be supported by the Arrowhead Framework, which will hide most of the underlying communication which most often consumes a considerable amount of engineering time.

If the partitioning is made with engineering in mind, the number of real-time closed loops will be limited. Then the engineering will have a limited number of communication links to consider to the latency prediction and also a limited number of computations to be handled on the device or devices involved in controller computations. Having limited local clouds reduces engineering time regarding latency and computation time verification.

The local cloud provides a certain number of features like

- cloud service infrastructure

- automation support services

- predictable latency properties - well-defined usage of communication physical layer and lower layer protocols

- security properties - authentication and authorisation

Each of these features can be used by a plant control and automation design engineer to build the desired application services and the auxiliary support systems needed by the control application, thus reducing the complexity of the local cloud design even further.

A further support to plant automation design needed is the capturing of machinery involved and their functionality and interaction.

2.6 Latency in local clouds

As previously stated, automation has strong requirements on real-time data exchange and control. Here the communication latency in a service exchange becomes critical. The total latency is dependent on communication channel bandwidth, protocol packet overhead, and the size of the data transmitted. The chosen technology bandwith defines a lower limit to real-time resolution obtainable. This is further reduced by the payload of data and the protocol overhead.

For certain communication technologies there are possibilites to state latency guarantees. Most often the communication is based on TDMA medium access approach [10, 11]. In TDMA the available communication bandwith is divided into a number of timeslots. Each timeslot can then be assigned to a certain actor or activity in the local cloud, thus enabling differentiated latency within the local cloud. Some examples of real-time communication solutions based on TDMA are

- IEEE 802.15.4 (TDMA mode) [12]

- Time Triggered Protocol [13]

- Industrial Ethernet [14]

- Fieldbus/Profibus [15, 16]

On-going research addresses improvement in real-time performance on both wireless and wired communication. When packet losses disturbes a TDMA timeslot, the latency guarantee cannot be handled by TDMA. Such risk is obviously higher for wireless links than for wired links. This an on-going field of research. For further discussion on this topic, see, for example [17, 18, 19].

The payload and protocol size is the other factor that will determine latency over a network link. Here payload compression is established as the means of reducing service payloads. The payloads are most often encoded in XML or JSON [20, 21]. EXI is a standardised approach to compress XML or JSON encoded payloads [22, 23]. Compression rates of about 30 can be expected [24]. This will clearly help to reduce the network link latency for a service transfer. On the cost side comes the compression and de-compression computation time. This is of course dependent on the CPU speed and memory available. In 2012 Kyusakov et.al. [24] reported compression/de-compression times averaging 15 ms on a basic PC. With the improvement to be expected over time on processor speed code compression seems to provide a good contribution to enabling bindings of link latency for service data exchanges.

By specifying the communication hardware and MAC layers to be allowed in a local cloud in combination with compression, there are measures to provide certain real-time guaranties within a specific local cloud. The minimum latencies achievable will be dependent on number of nodes, type of MAC, type of transport protocol, payload compression, etc. Network simulation tools like NS-3 [25] enable the prediction of latency in a specific local cloud setting.

2.7   Security in local clouds

One of the arguments for local clouds is the possibility to provide a local security "fencing" around the local cloud and appropriate authenticating, authorisation, and encryption of services within the local cloud. Such "fencing" does need ways of keeping non-internal network activity outside the local cloud. Thus a secure local cloud have should have firewalls at the interface to other clouds which support blocking of "external" traffic to come into the local cloud.

Such security fencing is dependent on control and minimising of communication in and out of the local cloud. This will reduce the number of holes made through the firewall in the local cloud security fence. The provision of all the administrative functionalities for a SOA-based system of systems within the local cloud is thus an important design principle. Thus there is no need for external resources in establishing the local cloud and its SOA-based infrastructure, as discussed in Section 2.2. This requirement is supported by the idea of having the mandatory core services deployed within a local cloud.

2.7.1 Service authentication and authorisation

Authentication is provided to ensure the identity of a service consumer requesting access to a provided service. The authorisation is to ensure that the requesting system has the right to consume the requested service.

For IoT-based automation, a local cloud may hold both resource-constrained devices and more powerful devices. They have different capabilities in handling authentication and authorisation technologies while still performing its dedicated automation task. Thus a local cloud should be capable of having one or a combination of authentication and authorisation methods like

- Strong certificates like X.509 [26, 27]

- Ticket-based solutions like Radius or Kerberos [28, 29]

The distribution of certificates is a task on its own. There are two different ways of getting a certificate:

- create one yourself (using the right tools, such as keytool [30]), or

- ask a Certification Authority to issue you one (either directly or using a tool such as keytool to generate the request).

In the case of tickets, the generation and distribution are made by a local, e.g., radius or kerberos server [31, 32].

In a local cloud a system has to be authorised to access a service. Initially it is assumed that a system owns the authority to deny or allow any service access of its own services. When a system tries to register its services to a local cloud, the local authentication system has to authorise such registration. The information on which systems that are expected to register services to the local cloud is most often provided by the plant description service, which in turn get the information from a plant engineering tool.

If a system is authorised to register its service to the local clouds the service may have its own predefined authorisation database or the authorisation process can be handed over to the authorisation system within that local cloud. In most cases foreseen, the authorisation should be handed over to the local authorisation system.

2.7.2   Data encryption

Data encryption can be provided either by enforcing IP communication encryption or relying on SOA protocol provided payload encryption.

IPSec [33, 34] is an end-to-end security scheme operating in the Internet Layer of the Internet Protocol Suite. IPSec protects all application traffic over an IP network. Applications can be automatically secured by IPSec at the IP layer. The downside of IPSec is the computational efforts required by each node in the network to handel IPSec key updates which are recommended to be every 20 minutes.

Other Internet security systems in widespread use, such as Transport Layer Security (TLS) and Secure Shell (SSH), operate in the upper layers at the Application layer. Some SOA protocols like CoAP or MQTT use TLS or DTLS [35, 36, 37, 38] within the protocol, thus protecting the payload during transport.

Local clouds should allow for the use of a mixture of these encryption approaches. However, if an application has need for services using different encryption technologies, a translation between different encryptions may be needed. This might open up man-in-the-middle security holes.

2.8   System of Systems scalability

With the number of IoT devices and their expected use in different automation tasks supporting smart city and smart grid concepts and the expected digital integration between different stakeholders there is a clear need for digitisation and automation solutions that will scale beyond the size of a local cloud and beyond the size of current automation systems without having nonlinear increases in engineering costs and nonlinear increases in possible security problems.

The building of larger digitisation and automation systems will in many situations call for information which for certain reasons is not available within a specific local cloud. For this purpose local clouds should be capable of inter-cloud interaction while maintaining SOA capabilities. Following the SOA based IoT and SoS approach there is a need for:

- Service discovery external to the local cloud

- Orchestration of service exchanges between systems residing in more than one local cloud

- Handling of security, authentication, authorisation, and data encryption for service exchanges between systems residing in more than one local cloud

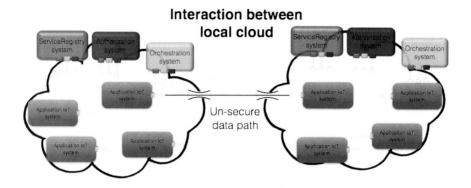

FIGURE 2.6
Inter-cloud service orchestration exchange diagram.

- Service exchange between systems residing in two different local clouds

- Protocol, encoding, and semantics translation supporting service exchanges between systems applying different SOA, protocols, encodings, and semantics

The properties of inter-cloud service exchanges need to be provided without sacrificing cloud internal security or risking impairment of network latency or simplified automation engineering.

2.8.1 Inter-cloud service exchange security

The need for inter-cloud communication opens up security challenges. Every external communication is punching a hole in the security fence created by a firewall at the rim of the local cloud. Every such hole should, from a security point of view, be regarded as a risk. For the inter-cloud service exchange there are two processes that will punch holes in the local cloud security fence. These processes are

- Orchestration of an inter-cloud service exchange

- The service exchange data path

Let's consider an orchestration scenario as in Figure 2.6. Here we find the following service exchanges that will require a hole in the local cloud security fence:

- Service discovery

- Obtaining orchestration information (service exchange end points).

- Authentication and authorisation of the service exchange.

The local cloud objective is to

- provide a way of making service discovery external to the local cloud where the discovery process is initiated

- enable orchestration push and pull distribution of service exchange end points between systems residing in different local clouds

- enable system authentication and service consumption authorisation between systems residing in different local clouds

In this way the local cloud concept with inter-cloud communication and interaction seams to have a number of interesting properties with regards to scaling.

Bibliography

[1] H. Derhamy, J. Eliasson, J. Delsing, and P. Priller, "A survey of commercial frameworks for the internet of things," in *Emerging Technologies & Factory Automation (ETFA), 2015 IEEE 20th Conference on*. IEEE, 2015, pp. 1–8.

[2] T. Erl, *SOA Principles of Service Design (The Prentice Hall Service-Oriented Computing Series from Thomas Erl)*. Upper Saddle River, NJ, USA: Prentice Hall PTR, 2007.

[3] "Arrowhead project." [Online]. Available: http://www.arrowhead.eu

[4] B. Scholten, *The Road to Integration: A Guide to Applying the ISA-95 Standard in Manufacturing*. ISA, 2007.

[5] (2016). [Online]. Available: http://isa99.isa.org/ISA99%20Wiki/Home.aspx

[6] "Iso/iec 81346 industrial systems, installations and equipment and industrial products – structuring principles and reference designations," ISO/IEC, Tech. Rep., 2009.

[7] "Iec 61360 - common data dictionary," IEC, International standard CDD - V2.0013.0002, 2005.

[8] Wikipedia, "Caex — wikipedia, the free encyclopedia," 2015, [Online; accessed 10-April-2016]. [Online]. Available: https://en.wikipedia.org/w/index.php?title=CAEX&oldid=681828591

[9] P. Adolphs and U. Epple, "Status report: Rami4.0," VDI/VDE-Gesellshact Mess- und Automatisierungstechnik, Tech. Rep., June 2015.

[10] R. Nelson and L. Kleinrock, "Spatial tdma: A collision-free multi-hop channel access protocol," *Communications, IEEE Transactions on*, vol. 33, no. 9, pp. 934–944, 1985.

[11] K. S. J. Pister and L. Doherty, "Tsmp: Time synchronized mesh protocol," in *Proceedings of the IASTED International Symposium on Distributed Sensor Networks*, 2008.

[12] "Wirelesshart specification 75: Tdma data-link layer, rev. 1.1." HART Communication Foundation Std., Tech. Rep. hCF_SPEC-75, 2008.

[13] H. Kopetz and G. Bauer, "The time-triggered architecture," *Proceedings of the IEEE*, vol. 91, no. 1, pp. 112–126, 2003.

[14] T. Skeie, S. Johannessen, and Ø. Holmeide, "Timeliness of real-time ip communication in switched industrial ethernet networks," *Industrial Informatics, IEEE Transactions on*, vol. 2, no. 1, pp. 25–39, 2006.

[15] E. Tovar and F. Vasques, "Real-time fieldbus communications using profibus networks," *Industrial Electronics, IEEE Transactions on*, vol. 46, no. 6, pp. 1241–1251, 1999.

[16] D. A. Glanzer, "Hse: An open, high-speed solution for plantwide automation," Fieldbus foundation, Tech. Rep., 2016.

[17] F. Dobslaw, "End-to-end quality of service guarantees for wireless sensor networks," No. 234, Mid Sweden University, 2015.

[18] E. Toscano and L. L. Bello, "Multichannel superframe scheduling for ieee 802.15. 4 industrial wireless sensor networks," *IEEE Transactions on Industrial Informatics*, vol. 8, no. 2, pp. 337–350, 2012.

[19] W. Shen, T. Zhang, F. Barac, and M. Gidlund, "Priori- tymac: A priority-enhanced mac protocol for critical traffic in industrial wireless sensor and actuator networks," *IEEE Transactions on Industrial Informatics*, vol. 10, no. 1, 2014.

[20] (2016) Extensible markup language — xml. [Online]. Available: https://en.wikipedia.org/wiki/XML

[21] (2016) Introducing json. [Online]. Available: http://www.json.org

[22] D. Peintner and S. Pericas-Geertsen, "Efficient xml interchange (exi) primer," W3C, Tech. Rep., 2014.

[23] R. Kyusakov, P. P. Pereira, J. Eliasson, and J. Delsing, "Exip: a framework for embedded web development," *ACM Transactions on the Web (TWEB)*, vol. 8, no. 4, p. 23, 2014.

[24] R. Kyusakov, J. Eliasson, and J. Delsing, *Efficient structured data processing for web service enabled shop floor devices*. IEEE, 2011, pp. 1716–1721.

[25] T. R. Henderson, M. Lacage, G. F. Riley, C. Dowell, and J. Kopena, "Network simulations with the ns-3 simulator," *SIGCOMM Demonstration*, vol. 14, 2008.

[26] M. Thompson, W. Johnston, S. Mudumbai, G. Hoo, K. Jackson, and A. Essiari, "Certificate-based access control for widely distributed resources," in *Proc. 8th UsenixSecurity Symposium*, 1999.

[27] M. Myers, R. Ankney, A. Malpani, S. Galperin, and C. Adams, "X. 509 internet public key infrastructure online certificate status protocol-ocsp," RFC 2560, Tech. Rep., 1999.

[28] B. C. Neuman and T. Ts' O, "Kerberos: An authentication service for computer networks," *Communications Magazine, IEEE*, vol. 32, no. 9, pp. 33–38, 1994.

[29] A. DeKok and A. Lior, "Remote authentication dial in user service (radius) protocol extensions," RFC 6929, April, Tech. Rep., 2013.

[30] Oracle. (2016) keytool - key and certificate management tool. [Online]. Available: http://docs.oracle.com/javase/6/docs/technotes/tools/windows/keytool.html

[31] "The freeradius project." [Online]. Available: http://freeradius.org

[32] MIT, "Kerberos: The network authentication protocol." [Online]. Available: http://web.mit.edu/kerberos/

[33] N. Doraswamy and D. Harkins, *IPSec: the New Security Standard for the Internet, Intranets, and Virtual Private Networks*. Prentice Hall Professional, 2003.

[34] V. Devarapalli and F. Dupont, "Mobile ipv6 operation with ikev2 and the revised ipsec architecture," IETF, Tech. Rep. RFC 4877, 2007.

[35] D. McGrew and E. Rescorla, "Rfc5764 — datagram transport layer security (dtls) extension to establish keys for the secure real-time transport protocol (srtp)," IETF, Tech. Rep., 2010.

[36] E. Rescorla and N. Modadugu, "Datagram transport layer security rfc - rfc 4347," IETF, Tech. Rep., 2006.

[37] Z. Shelby, "The constrained application protocol (coap) - rfc 7252," IETF, Tech. Rep., 2014.

[38] (2016) Mqtt is a machine-to-machine (m2m)/"internet of things" connectivity protocol. [Online]. Available: http://mqtt.org

3

The Arrowhead Framework architecture

Jerker Delsing
Luleå University of Technology

Pal Varga
AITIA Inc

Luis Ferreira
ISEP, Polytechnic Institute of Porto

Michele Albano
ISEP, Polytechnic Institute of Porto

Pablo Puñal Pereira
Luleå University of Technology

Jens Eliasson
Luleå University of Technology

Oscar Carlsson
Midroc Automation

Hasan Derhamy
Luleå University of Technology

CONTENTS

3.1   Architecture fundamentals

The objective of the Arrowhead Framework architecture is to facilitate the creation of local automation clouds. Thus enabling local real time performance and security, paired with simple and cheap engineering, while simultaneously enabling scalability through multi-cloud interaction.

The architecture addresses the move from large monolithic organisations towards multi-stakeholder cooperation where cooperation is fostered by market requirements. This is to support the high-level topics in today's society such as sustainability, flexibility, efficiency, and competitiveness in production.

Devices in such local clouds are considered to be IoT devices speaking at least one SOA (Service-Oriented Architecture) protocol. The capability of building automation systems requires a number of local cloud properties to be enabled. Furthermore, both intra- and inter-cloud information service exchange capabilities are necessary for enabling IoT devices to inter-operate and to be integrated with others to become an automation System of Systems. Previous work in this field comes from several larger EU projects such as Socrades and IMC-AESOP. The Arrowhead Framework architecture is building on the results of these projects [1, 2].

To facilitate this objective, the Service-Oriented Architecture paradigm is used. Thus the following properties are important starting points:

- Loose coupling

    - Autonomy - a service exchange is not supervised
    - Distributed - services are distributed over several devices
    - A system is responsible, owns the information, and can decide whom to share with

- Late binding

    - Possible to use information any time by connecting to the correct resource at a given time

- Lookup

    - Publish and register services to notify others about endpoints (how to reach me)
    - Discover others that I comply with (expected/wanted ServiceType)

The design of the Arrowhead Framework is further based on the following fundamentals:

- A system producing a service has the initial authority of its own service offering

- Information assurance shall be at the service exchange level

- Information centric networking

The Arrowhead Framework allows for

- A Publish-Subscribe approach

- Both the Push and Pull approaches

- Dynamic creation of new services and their subsequent usage

The above properties, fundamentals, and functionalities are provided by the Arrowhead Framework through

- A minimal set of mandatory services to create a System of Systems

- A set of automation support services - facilitating design of application System of Systems

A developer needs to know how to develop, deploy, maintain, and manage Arrowhead compliant systems. Therefore, it is crucial that there is a common understanding of how Arrowhead Framework services, systems, and System of Systems are defined and described. To address these issues, the framework also includes design patterns, documentation templates and guidelines that aim at helping systems, newly developed or legacy, to conform to Arrowhead Framework specifications.

In the following we will discuss the Arrowhead Framework local automation cloud architecture. The architecture and a number of the core services implementing the architecture are open source. The Arrowhead Framework community development site is located at `http://forge.soa4d.org/plugins/mediawiki/wiki/arrowhead-f/index.php/Main_Page` also reachable from `http://www.arrowhead.eu`.

3.2   Important definitions

To discuss and define a local cloud architecture, a few definitions are important. The Arrowhead Framework local automation cloud architecture makes use of the following important key words:

- Service

- System

- Device

- Local cloud

- System of Systems

These key words may have other definitions in different domains and contexts. But, in this book and within the Arrowhead Framework the following definitions are used.

3.2.1   Service

In the context of the Arrowhead Framework, a service is what used to exchange information from a providing system to a consuming system, see Figure 3.1. In a service, capabilities are grouped together if they share the same context [3]. A service can be implemented to use a number of different SOA protocols. Some examples of SOA protocols are REST [4], COAP [5], XMPP [6], MQTT [7, 8], or OPC-UA [9].

A service is produced by a software system (see below). A service can have associated meta-data and can be capable of supporting non-functional

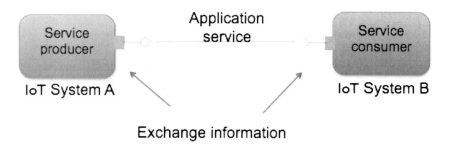

FIGURE 3.1
Services provide information exchange between a service-producing system and a service-consuming system.

requirements such as security, real-time operation, or different levels of reliability among others.

It must be possible for an Arrowhead Framework compliant service to

- be registered with the Arrowhead Framework mandatory core systems

- be consumed or provided by an Arrowhead Framework compliant system

    A service may be capable of

- being dynamically configured

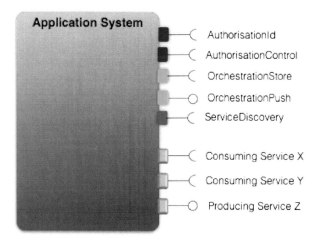

FIGURE 3.2
A system is capable of consuming the Arrowhead Framework mandatory core services and will produce and/or consume one or more services.

3.2.2   System

An Arrowhead Framework system is what is providing and/or consuming services, see Figure 3.2. A system can be the service provider of one or more services and at the same time the service consumer of one or more services. A system is implemented in software and executed on a device (see below). A system can have associated meta-data. We are here separating the software based system from the hardware - device. The reason is grounded in security considerations. To achieve a chain of trust, a piece of "computing" hardware need to be identifiable. The same is required for a software application executing on that hardware. In the Arrowhead context an software application capable of producing and/or consuming services is named system and the "computing" hardware hosting a system is named device.

An Arrowhead Framework compliant system shall be capable of:

- consuming Arrowhead Framework compliant services

- producing Arrowhead Framework compliant services

- registering itself to the mandatory SystemRegistry

- releasing its authority of its own service offering to a local cloud orchestration system

  A system may be capable of

- creating new services on demand

- being dynamically configured

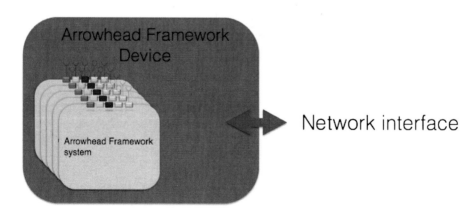

FIGURE 3.3
An Arrowhead Framework device is a equipment capable of hosting systems exchanging services.

3.2.3 Device

An Arrowhead compliant device is a piece of equipment, machine, hardware, etc. with computational, memory and communication capabilities which hosts one or several Arrowhead Framework systems and can be bootstrapped in an Arrowhead Framework local cloud, cf. Figure 3.3. Any other device, equipment, machine, hardware, component etc. is non-Arrowhead compliant.

A device may be capable of

- dynamically hosting new systems and their services

- being registered to the DeviceRegistry

- being dynamically configured

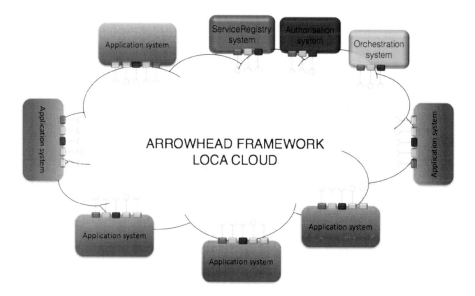

FIGURE 3.4
An Arrowhead Framework local cloud hosts at a minimum the three mandatory core services, ServiceRegistry, Orchestration, and Authorisation and one or more application systems.

3.2.4 Local cloud

In the Arrowhead Framework context a local cloud is defined as a self-contained network with the three mandatory core systems deployed and at least one application system deployed, cf. Figure 3.4. A local cloud shall host only one ServiceRegistry system. For administrative and security reasons it is strongly recommended that only one instance of the other two mandatory core services are deployed in a local cloud.

It is advisable that a local cloud holds a mean of distributing IP addresses to joining devices. It is further advisable that the local cloud has firewall protection to surrounding networks. By which external network traffic can be blocked from reaching the interior of the local cloud.

3.2.5   System of Systems

A System of Systems within the Arrowhead Framework is defined as a set of systems, which are administrated by the Arrowhead mandatory core systems and exchange information by means of services.

A local cloud thus becomes a System of Systems in the Arrowhead Framework's definition. If two systems reside in different local clouds that are administrated by Arrowhead core systems to exchange services, it also is a System of Systems in the Arrowhead Framework's definition.

When Arrowhead compliant systems collaborate, they become a System of Systems. Since two or more such Systems of Systems can collaborate, the Arrowhead Framework becomes a natural enabler of further, complex solutions. Figure 3.5 depicts such an example.

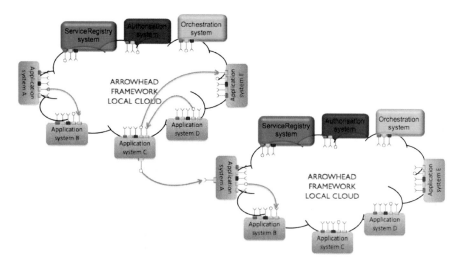

FIGURE 3.5

Arrowhead Framework Systems exchanging services, thus building a system of systems within a local cloud and between local clouds.

Service exchanges between systems can be initiated by an orchestrating system providing orchestration rules to the involved systems. Service exchanges can also be initiated by apriori knowledge of one system to seek a specific service via a ServiceRegistry. Upon a seek criteria match a service exchange can be initiated. To support a structured governance of a System of Systems, the preferred approach is the use of an Orchestration system.

3.2.6 Service, system, device and local cloud identifiers

In order to allow service discovery, system administration, and device mappings, there is a need to uniquely define identifiers to the devices, systems, and services. Thus, to enable proper identification of which device and system a specific service is executed by, there is a need for a way of identifying

- Services

- Systems

- Devices

- Local cloud

Here references are made to the SysD, SD, IDD, CP and SP documents. For definitions of these documents please see Section 3.3 below.

There are two classes of Arrowhead Framework services:

- Core services, sub-divided into

  - Mandatory core services

  - Automation support services

- Application services

The identification of Arrowhead Framework services follows the same idea as identifiers in DNS-SD records [10, 11, 12]. Thus an Arrowhead Framework service is identified using the following structure, which follows the RFC-6335 recommendations [13]:

- Core services
  _ServiceName._ahfc-ServiceType._protocol._transport.domain

- Application services
  _ServiceName._ahf-ServiceType._protocol._transport.domain

- *.domain* is the domain name under which the device with an associated system has an IP address, e.g., *subdomain.domain.topdomain.* An example of a domain name can be *app.arrowhead.eu.*

The detailed explanation is

- *_ServiceNames* is the name of the particular service instance, e.g., *_Temp102.*

- *_ahf/c-servicetype*: The Arrowhead Framework makes use of the selective service instance enumeration (subtypes) possibility as described in RFC-6763 [10]. For core services the ServiceType always starts with *_ahfc-* and *_ahf-* is reserved for Arrowhead Framework application services.

- – *_ahfc-ServiceType* is an Arrowhead Framework core services type where e.g. the service type orchestration is added, leading to the complete ServiceType of *_ahfc-orchestration*. The above specification should be given in the SD (Service Description) document.

- – *_ahf-ServieType* is an Arrowhead Framework application services type. Here e.g. vibration is the service type, leading to the complete ServiceType of *_ahf-vibration*. Above specification should be given in the SD document.

- *_protocol* is, e.g., *_coap* when the CoAP [5] protocol is used and specified in the IDD (Interface Design Description) document.

- *_transport* is, e.g., *_udp* when the UDP [14, 15] transport protocol is used and specified in the IDD document.

The end point to an application service instance consists of a path and a port. Using, for example, REST [4], an end point may look like *http://app.arrowhead.eu/ahf/Temp1/8090/*.

To enable discovery of interoperability at the service level, there is a need for information on payload encoding, compression and semantics. In the context of the Arrowhead Framework, this is regarded as service meta-data. Service meta-data is to be provided through the DNS TXT record using the following key pairs:

- *encode=syntax*, e.g., *encode=xml* when XML [16] encoding is used and specified in the CP (Communication Profile) document.

- *compress=algorithm*, e.g., *compress=exi* when EXI [17] compression is used and specified in the CP document

- *semantic=XX*, e.g., *semantic=senml* when SenML [18] semantics is used and specified in the SP (Semantic Profile) document.

A device is identified through a DeviceName plus the associated MAC address of its network interface. Hereto an ID key shall/may be associated using a secure bootstrap process; see Section 3.4.4. This enables us to associate the device hardware and its network interface to its IP address and MAC address, allowing for building service exchange topology information useful for quality of service management. For a device with more than one network interface device instances with the same DeviceName but different MAC addresses should be defined. The above information shall be specified in the SysD (System Description) document.

An Arrowhead Framework software system is identified through a system ID key generated from a two-way asynchronous authorisation process involving a trusted part and the device name instance identifying which device hardware is hosting the system and which network interface the software system is using. This shall be specified in the SysD document.

A local cloud is identified with the servicename and service type of the ServiceRegistry system. An example is _ *cloud-x._ahfc-ServiceDiscovery.*

The following four service types are considered to be well known:

- _ *ahfc-ServiceDiscovery*

- _ *ahfc-SystemDiscovery*

- _ *ahfc-DeviceDiscovery*

- _ *ahfc-AuthorisationControl*

This allows for simple deployment with authorisation registration of devices, systems and services with the local cloud.

3.3   Documentation structure

It is a common experience of both system developers and integrators that insufficient and not properly structured documentation makes it hard – if not impossible – to properly understand how to integrate with a given system.

A system integrator needs to know how to develop, deploy, maintain, and manage Arrowhead compliant systems. To address these issues, the Arrowhead Framework document structure allows documenting SOA artefacts in a common format [19].

For this purpose the Arrowhead consortium [20] defined a three-level documentation structure: System of Systems, system, and service level. These are depicted in Figure 3.6, which also shows the links between documents.

The main concept of the documentation structure is to provide abstract and implementation views of the Systems of Systems, systems, and services. The main purpose of the abstract view documents is to allow any developer to implement System of Systems, systems and services, based on these documents. This further support that the resulting implementation becomes Arrowhead Framework compliant.

3.3.1   System of Systems level documentation

The System of Systems level consists of two types of documents. The System of Systems Description (SoSD), which shows an abstract view of an SoS, and the System of Systems Design Description (SoSDD), which shows an implementation view of the SoS with its technologies and deployment views.

The SoSD describes the main functionalities and the generic architecture of the SoS. It will mainly be used to describe one System of Systems in an abstract way, without instantiating into any specific technologies. Examples of its usage are the description of generic SOA-based installations, like building

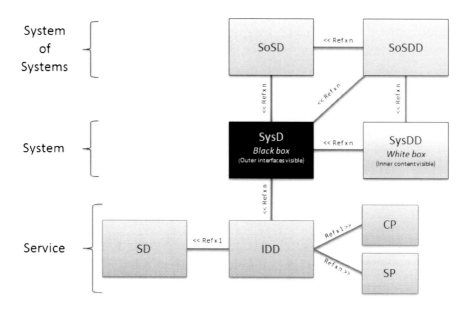

FIGURE 3.6
Arrowhead Framework documentation relationships.

automation systems or a factory automation system. The document should present its main building blocks as independent systems with pointers to their specific abstract view documents, the system Descriptions (SysDs) see below. Also, diagrams representing system behaviour, like use-case diagrams and behaviour diagrams (e.g., using UML, BPMN, SysML, AutomationML [21, 22, 23, 24]) must be included. This document also includes information about non-functional requirements, like required levels of QoS and security.

For the Arrowhead Framework itself there is a generic SoSD for which the current version can be found at the Arrowhead Framework wiki [25].

The SoSDD document describes how an SoSD has been implemented on a specific scenario, showing the technologies used and its setup. Therefore, it points out all necessary black box SysD and white box SysDD documents, describing the systems used in this realisation. The SoSDD should also contain behaviour diagrams which clearly identify the technologies used and the setup of this SoS realisation. The document can optionally include a description of its physical implementation and the non-functional requirements implemented by this realisation.

3.3.2 System level documentation

System level documentation consists of a black box System Description (SysD) document and a white box System Design Description (SysDD) document.

The SysD describes the system as a black box, documenting the system functionality and its hosted services and their provided and required interfaces with the corresponding technical solutions, without describing its internal implementation. The by the system provided service interfaces are referenced and defined in the Interface Design Description (IDD) document; see Section 3.3.3 below. The services provided are defined in the Service Design (SD) document, see Section 3.3.3 below. The SD document shall provide a clear definition of how to interface the system, thus enabling coding of a consumer system.

The SysDD extends the black box description, showing its internal details. This document is optional, since it might expose knowledge of the company which implemented the system, but it can be used as an internal document for future reference by the system owner.

3.3.3 Service level documentation

Service level documentation consists of four documents: the Service Description (SD), the Interface Design Description (IDD), the Communication Profile (CP), and the Semantic Profile (SP).

The IDD is pointed to by a SysD document. It states the actual implemented solution of a system. Here are defined the service identifiers of the specific service implementations. For the SOA protocol and encoding used the IDD is making reference to the Communication Profile (CP) document, see below. For data and information semantics the IDD make reference to the Semantics Profile (SP) document; see below.

The SD is a technology independent and abstract view of a service. The document describes the main objectives and functionalities of the service and its abstract interfaces. Further, an abstract information model shall be provided. Sequence Diagrams showing how the service is interacted with, shall also be provided.

The CP contains all the information regarding the transfer protocol, data compression, data encryption, and data encoding used, e.g., CoAP, UDP, EXI, DTLS, and XML.

The SP defines the data and information semantics used, e.g., SenML.

3.4   Arrowhead Framework architecture

Based on the SOA fundamentals and principles discussed previously, a local cloud will require three fundamental properties:

- Capability to register a service to the local cloud

- To discover which services are registered with the local cloud

- Enabling loosely coupled data exchange between producer and consumer systems - orchestrate service exchanges

- Authentication of consuming systems and granting Authorisation of service exchanges

Following the above fundamentals, definitions, and documentation structures, a local automation cloud architecture is defined. The architecture is composed of a number of systems, which provide a number of services. The objective is an architecture from which self-contained local automation clouds can be created. These clouds shall further be capable of providing certain automation support services, and provide support for bootstrapping, security, suitable metadata, protocol and semantics transparency, and inter-cloud service exchanges.

For that purpose the architecture features three types of services:

- Mandatory core services

- Automation support core services

- Application services

These services are provided by mandatory and support core systems, as well as application systems.

The use of this set of mandatory core systems and services makes it possible to design and implement a minimal local automation cloud. These mandatory core services will enable the desired basic properties of a local cloud, as discussed in Chapter 2. In order to support important local cloud properties, a number of automation support core services are defined as part of the architecture. These core services are provided by core systems. Figure 3.7 summarises the core systems currently defined with the Arrowhead Framework. Utilising all of these core systems in a local cloud is not mandatory. Implementations and documentation of the mandatory core systems and many of the automation support core systems are available as open sources via the Arrowhead Framework wiki [25].

In the following sections the core systems providing core services are described at a system level, corresponding to the SysD document of the Arrowhead Framework documentation structure.

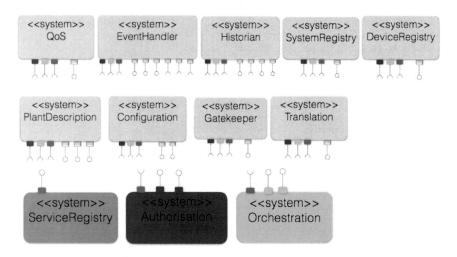

FIGURE 3.7
Core systems of the Arrowhead framework.

3.4.1 The mandatory core systems

These mandatory core services support the following fundamental properties assigned to a local cloud:

- Loose coupling
  - Autonomy - a service exchange is not supervised
  - Distributed - systems exchange services are distributed on several devices.
  - A system is responsible and owns its information and decides with whom to share
- Late binding
  - Possible to use information any time by connecting to the correct resource at a given time
- Lookup
  - Publish and register to notify others about endpoint (how to reach me)
  - Discover others that I comply with (expected/wanted service type)
- Information assurance at service exchange level
- Minimal set of mandatory services to create a System of Systems

The mandatory core systems and their service provide these fundamental properties to a local cloud. Thus enabling and allowing service exchanges between a service producer and a service consumer with desired level of security and autonomy, which is briefly illustrated in Figure 3.8 .

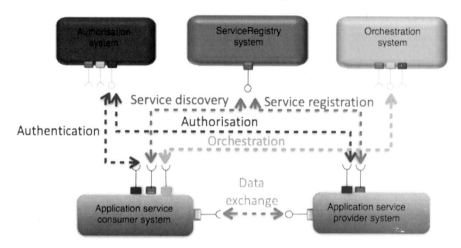

FIGURE 3.8
Minimal local cloud with indications of mandatory core service interaction enabling service exchange between two application systems.

Therefore, the creation of a minimal working local cloud based on the Arrowhead Framework must be based on three mandatory core services. The mandatory core services and their three hosting systems are

- ServiceRegistry system providing the

  - ServiceDiscovery service

- Authorization system providing the

  - AuthorisationControl service
  - AuthorisationManagement service
  - AuthenticationID service (only for ticket-based implementation)

- Orchestration system providing the

  - Orchestration service
  - OrchestrationManagement service

The objectives of these mandatory core systems with their system definitions and high-level descriptions are given below. In Chapter 4 the Arrowhead Framework implementation of the core systems and services is provided in some detail. For the current status of these implementations, please refer to the Arrowhead Framework wiki [25].

FIGURE 3.9
The mandatory ServiceRegistry system.

3.4.1.1   ServiceRegistry system (mandatory)

The objective of the ServiceRegistry system is to provide storage of all active services registered within a local cloud and enable the discovery of them.

The ServiceRegistry system keeps track of all active services produced within a local cloud. It provides a service registry functionality based on DNS and DNS-SD standards [10, 11, 12], since the Arrowhead Framework is a domain-based infrastructure. The ServiceRegistry is an independent system that provides one service and does not consume any other services, cf. Figure 3.9. All ServiceRegistry system graphs are color coded in blue.

All systems within the local cloud with services producing information to the local cloud shall publish their service within the ServiceRegistry by using the ServiceDiscovery service. Using the DNS-SD engine, in the ServiceRegistry system, services can be published, un-published or looked up. The ServiceRegistry system holds all published and active services within the local cloud. A cleaning mechanism for broken services is still to be defined. The usage of the DNS-SD for storing registered services is the basis for the service identifier structure described in Section 3.2.6.

All systems within the local cloud with services that produces information shall publish their producing service with the ServiceRegistry system by using the ServiceDiscovery service. If a system disconnects from the local cloud, its services shall be de-registered in the ServiceRegistry system. To address broken systems and network failures a registration time out and keep alive approach is recommended. The DNS PTR and DNS SRV, TimeToLive, TTL, record provide such time out data. Thus a system should re-register its services according to this TTL data. If not done the ServiceRegistry can unpublish the service.

The details of a service layer agrement, SLA, holds information on service protocol, transport protocol, interface, associated methods, datatypes, encoding, semantics, and compression. There are several approaches to provide this information. The obvious one is from design time documentation. The more attractive approach is to provide such information as meta-data with the registered service. In this way the SLA provided by the service producer can be retrieved and decoded by a consumer.

There exist some technologies that support SLA description; examples are

WSDL [26], WADL [27] and HATEOAS [28]. In the ServiceRegistry the SLA information shall be provide as meta-data. Technically it's stored in the DNS TXT field as key pairs. The following key pairs indicates how the SLA can be provided to a service consumer:

- *wadl=link*

- *wsdl=link*

- *hateoas=link*

The ServiceRegistry in addition can hold information regrading priorities within a System of Systems. The current approach to this is through the priority and weight fields of the DNS SRV record. This information is intended for use by Orchestration systems and quality of service systems to support automation QoS and dynamic re-organisation of automation operations based on QoS and service availability. The details of this, still a matter of further investigations.

For operator interaction with the ServiceRegistry, an optional MMI-ServiceRegistry system is defined, see Figure 3.10. This system provides a graphical user interface enabling the listing of published services.

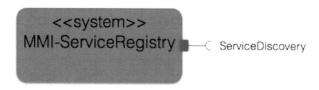

FIGURE 3.10
The optional MMI-ServiceRegistry system.

3.4.1.2   Authorization system (mandatory)

The objective of the Authorisation system is to provide Authentication, Authorisation and optionally Accounting of a system consuming a produced service. Based on the set of authorisation, authentication and optional accounting rules a service provider can ask whether a consumer is allowed to use a service resource or not.

Two different Authorisation systems are defined within the Arrowhead Framework. An AA - Authorisation Authentication system is defined. Next an AAA - Authorisation Authentication Accounting system is defined.

Based on existing technologies, the AA system is better suited for local clouds enrolling systems hosted on devices with sufficient computational power. While the AAA system is better suited for local clouds enrolling systems hosted on resource constrained devices. All Authorisation system graphs are color coded in red.

AA - Authorization system

The AA Authorisation system implements an Authorisation system based on X.509 certificates [29]. It requires some computation power from a device and is thus not suitable for very resource constrained devices.

The Authorisation system provides two services and consumes one service, see Figure 3.11. The Authorisation system provides the ability to define and check the access rule for the consumption of services and its resources. Based on the access rules the service providers can ask whether a consumer is allowed to consume the service resource or not.

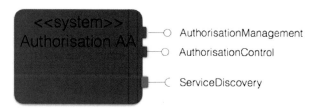

FIGURE 3.11
The mandatory Authorisation system, AA version.

The AuthorisationManagement service offers the possibility to manage the fine grained access rules for specific resources and also configure specific properties of the ticket like time-out. The AuthorisationControl service provides the possibility of controlling access to a service and a particular resource within the local cloud. This system includes both the authentication of the consuming system and the authorisation to consume a requested service. The system consumes the ServiceDiscovery service to publish the two produced services with the ServiceRegistry system.

AAA - Authorisation system

The AAA — Authorisation system implements an Authorisation system based on Radius tickets [30]. This solution is feasible to apply in local cloud hosting resource constrained devices.

The ticket based AAA — Authorization system provides three services and consume the ServiceDiscovery service, see Figure 3.12.

The AuthorizationManagement service offers the possibility to manage the fine grained access rules for specific resources and also configure specific properties of the ticket like time-out. The AuthorizationControl service provides the possibility of controlling the access to a service and a particular resource within the local cloud. This system includes both the authentication of the consuming system and the authorisation to consume a requested service.

In addition to the services provided by the AA system, an AuthenticationID service is provided by the AAA system. There is a challenge-response mechanism to get a valid ticket to authenticate each consumer on the local

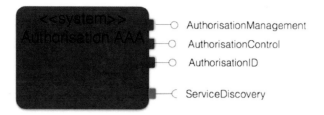

FIGURE 3.12
The mandatory ticket-based Authorisation system and its consumed and produced services.

cloud. The system consumes the ServiceDiscovery service to publish the three produced services with the ServiceRegistry system.

3.4.1.3   Orchestration system (mandatory)

The Orchestration system is a central component of the Arrowhead Framework and also in any SOA-based architecture [3]. In applications the use of SOA for a massive distributed System of Systems requires orchestration. It is utilised to dynamically allow the re-use of existing services and systems in order to create new services and functionality [31].

The process of orchestration is essential in support of service re-usability, service discoverability and service composability. From an architectural point of view, the Orchestration system is responsible for finding and pairing service consumers and providers. The input for this pairing is provided from, e.g., a plant description service, engineering tool, or an operator. In this regard, it provides advanced service discovery for systems. Orchestration is a key enabler for the engineering of Systems of Systems. Engineering information about suitable services, authorisation, and QoS is input to match-making algorithms and negotiation with the Authorisation and QoSManager systems.

The result of an orchestration request can vary over time, dependent on the environment and the requirements. This is because, in order to find the optimal service for the application system, there is much negotiation, with QoS for example, which may lead to different results. Therefore the orchestration must avoid oscillation in service selection. Additionally within the scope of an local cloud it is strongly recommended that only a single Orchestration system is deployed.

The Orchestration system stores orchestration requirements and resulting orchestration rules. The requirements are specified using engineering tools, and provided to the Orchestration system either via a PlantDescription service or directly. New requirements can be provided in run-time upon which the Orchestration system computes new orchestration rules which are provided to the involved application systems. These rules consist of the endpoints where the produced service of interest is found.

In a local cloud, orchestration provides service consuming systems with service consumption patterns and endpoint information of the produced services to be consumed. Based on this information, the consuming system can request to consume the assigned service in an autonomous manner until a new orchestration is pushed to or pulled by a consuming system.

Some systems may have the capability to dynamically and/or temporarily create new service providers. For this purpose such system shall produce a service type called _ahf-servprod/_ahfc-servprod. An Orchestration system shall be able to consume such _ahf-servprod/_ahfc-servprod types and thus request the creation of such new service instances. In some cases there is a need to dynamically create new service providers in order to fulfil the System of Systems requirements, for example, in the case of the Arrowhead translator. Thus, while performing matchmaking, the Orchestrator system is able to consume the translation's _ahf-servprod/_ahfc-servprod service and instantiate a new translator for injection into the service consumption pattern.

Thus the objective of the Orchestration system is to provide a mechanism for distributing orchestration rules and service consumption patterns.

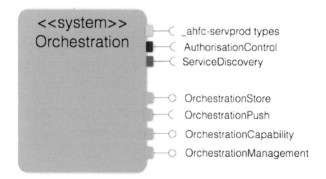

FIGURE 3.13
The orchestration system and its consumed and produced services.

The Orchestration system produces three services and consumes four services, cf. Figure 3.13. All Orchestration system graphs are color coded in green. The produced service, OrchestrationManagement, provides the possibility to manage the connection rules for specific services. Next the produced service, OrchestrationStore, provides the possibility for an application system to pull orchestration rules. Finally the produced service OrchestrationCapabilities give the number of consumers currently consuming a specific service and which producers and consumers that are available for a certain service type. The service Orchestration Capability can be used for orchestration and/or to create a current state picture over the complete Arrowhead local cloud system-to-system interactions and information exchanges currently active.

The Orchestration system publishes its presence to the local cloud by consuming the ServiceDiscovery service from the ServiceRegistry sys-

tem. It further shall be capable of consuming services of the type
_ahf-servprod/_ahfc-servprod which then allows it to instantiate new service providers based on requirements and availability.

3.4.2   Automation support core systems

To facilitate automation application design, engineering, and operation, the
Arrowhead Framework further contains a number of automation support services provided by the related automation support core systems. The objective
of the automation support core systems is to support:

- The implementation of "plant" automation. Here plant could be the infrastructure of, e.g., a car manufacturing plant, a mine, an infrastructure

- Housekeeping within the local cloud

- Security and bootstrapping of a local cloud

- Inter-cloud service exchange

- System and service interoperability

For the purpose there are currently ten automation support core systems
and their associated services defined. The systems currently defined are expected to address the architecture stated above. It's clear that additional
automation support core systems and services will be defined in the future.

The automation support systems currently defined are

- PlantDescription system

- Configuration system

- DeviceRegistry system

- SystemRegistry system

- EventHandler system

- QoSManager system

- Historian system

- Gatekeeper system

- Translation system

These support systems are described below at a level corresponding to
their respective SySD document. All support system graphs are color coded
in yellow.

3.4.2.1 PlantDescription system

The objective of the PlantDescription system is to provide a basic common understanding of the layout of a "plant" or "site", providing possibilities for actors with different interests and viewpoints to access their view of the same dataset provided by other sources.

Considering the different types of objects and relations that are expected to be present in a future Internet of Things (IoT) network, the concept of displaying different aspects of the same objects appears to be a useful solution to be able to present them all in an engineering tool.

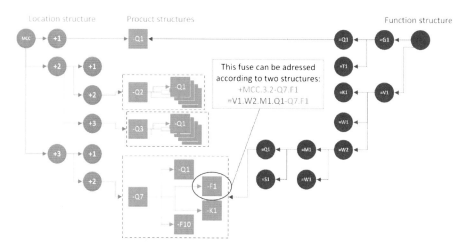

FIGURE 3.14

Example of how one component may be described by more than one viewpoint. Each colour in this diagram can be represented as a separate viewpoint by the plant description.

There are many different standards to describe automation objects. To keep a full database up to date would be an arduous if not impossible task, if the database is to contain all information that is required for all engineering tools and for all types of objects.

The approach of the PlantDescription system is instead to set up a separate system for identifying the objects and their relations while leaving the object details in the already established, standardised databases provided by existing engineering tools. Thus, throughout the design phase, all of the engineering data is still stored and maintained in the formats preferred by the engineering tools in their respective databases.

Figure 3.14 illustrates how a device or component documented according to IEC 81346 [32] may be found through traversing two different hierarchies, depending on the interest of the user. Each colour in the figure can be represented as a separate viewpoint by the plant description, with blue representing the location of the object, red representing the function, and green representing the components making up a product. This standard and its approach to identifying objects was a major point of inspiration for the design of the PlantDescription system.

Further discussion of the design and prototype implementation of a PlantDescription system is provided in [33, 34].

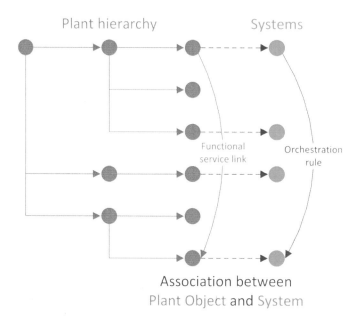

FIGURE 3.15
Illustration of how a relationship in the PlantDescription system can translate into an orchestration rule.

Figure 3.15 illustrates how a relationship between the logical objects in the PlantDescription system can be used to construct an orchestration rule, through the mapping of systems to plant objects. The "Functional service link" could, for example, be an internal link from an AutomationML structure, describing that the top object should send some data to the bottom object (where, e.g., the top object could be a sensor sending its data to a controller or actuator, represented by the bottom object). Such produced orchestration rules can then be provided to the Orchestration system via the OrchestrationManagement service.

The PlantDescription system produces three services and consumes the mandatory core services, as depicted in Figure 3.16.

The services produced are

- GetViewpoint

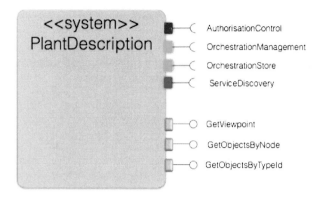

FIGURE 3.16
The PlantDescription system and its consumed and produced services.

The service is used with an argument, Type, to get all nodes and links that correspond to this type.

- GetObjectsByNode
  The service is used with a argument, NodeId, to get all nodes linked to the specified node, and those links.

- GetObjectsByTypeId
  The service is used with a combination of argument of Type and Node-TypeId corresponding to the identity by which the node is identified according to that type. The function returns the same information as the GetObjectsByNode for that node.

The services consumed are

- OrchestrationStore

- OrchestrationManagement

- ServiceDiscovery

- AuthorisationControl

3.4.2.2 Configuration system

The Configuration system allows systematic management of configurable systems. Depending on the nature of the System of Systems the Configuration system may be limited to a store from which an application system may get its updated configuration or send a backup of its current configuration. In other scenarios the Configuration system may be able to compile configuration files from engineering data provided and manage which application systems should update their configuration at which time.

The objective of the configuration system is to enable systematic management of configuration information to configurable systems. The Configuration system shall be able to store and backup application configuration information. Such configuration information shall be possible to pull or push from the Configuration system. Typical configuration scenarios are

- Deployment of an Arrowhead Framework compliant software system to a device

- Control code to PLCs and IoT controllers

- Configuration of sensors and actuators - e.g., sample rate, sensitivity, filtering . . .

- Configuration of HUIs

The Configuration system is producing one service and is consuming the mandatory core services, cf. Figure 3.17.

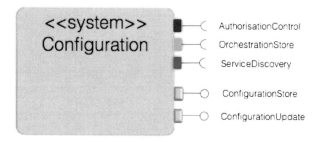

FIGURE 3.17
The Configuration system and its produced and consumed services.

The services produced are:

- ConfigurationStore

- ConfigurationUpdate

3.4.2.3 SystemRegistry system

The objective of the SystemRegistry system is to provide a local cloud storage holding the information on which systems are registered with a local cloud, meta-data of these registered systems and the services these systems are designed to consume.

The registration into a local cloud is part of the bootstrapping process of a local cloud, which is discussed in Section 3.4.4. The SystemRegistry system holds for the local cloud unique system identities for systems deployed within the Arrowhead Framework local cloud. This registry in combination with the

DeviceRegistry is necessary to create a chain of trust from a hardware device to a hosted software system and its associated services.

The SystemRegistry shall in addition to registering the system identity also store

- Metadata about the system
  addressing non-functional information such as software revision, deployment info, etc. A full definition is found in Section 4.3.3.

- Services the system is designed to consume
  which includes data on

  - ServiceTypes to consume: e.g., _ahf-pidcontrol_
  - SOA protocol capability: e.g., *http (REST)*
  - Transport protocol: e.g., *tcp or udp*
  - Payload data encoding: e.g., *JSON*
  - Payload semantics: e.g., *sensML*
  - Payload compression: e.g., *exi*
  - Service interfaces, methods, and datatypes supported, e.g., hard coded based on IDD-... or through, e.g., *WADL-link* or *HETAOES-link*

In the current definition, the SystemRegistry system is producing one service and is consuming the mandatory core services, cf. Figure 3.18.

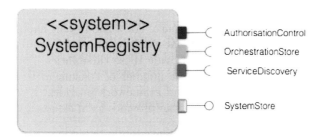

FIGURE 3.18
The SystemRegistry system and its produced and consumed services.

3.4.2.4 DeviceRegistry system

The objective of the DeviceRegistry system is to stores unique identities for devices deployed within an Arrowhead local cloud.

The DeviceRegistry system shall provide a local cloud storage holding information on which devices are registered with a local cloud. The registration into a local cloud is part of the bootstrapping process of a local cloud, which is

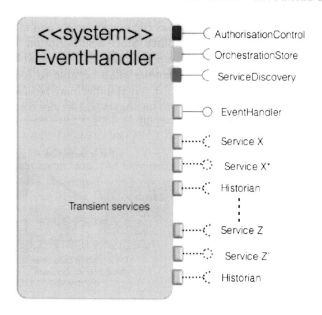

FIGURE 3.21

The EventHandler system and its produced and consumed services.

The EventHandler system produces the EventHandler service and consumes the mandatory core services, cf. Figure 3.21. The EventHandler service shall be consumed by the Orchestration system which initiates the creation of a transient pair of consumed and produced service X. To the consumed service X requested filtering and the subscription capability is applied. Which then is produced as service X_subscribe_filter. If requested the EventHandler also creates a transient service consumption of the Historian service, see Section 3.4.2.7.

3.4.2.6 QoSManager system

Quality of Service (QoS) within a local cloud is important. The automation requirements related to real-time communication and security have to be fulfilled. To achieve them, both monitoring of QoS and mitigation of QoS deviations shall be supported within a local cloud.

In the Arrowhead Framework architecture, the QoSManager system [36][37] will support QoS configuration and monitoring, in close collaboration with the Orchestration system.

Most of Arrowhead matchmaking between service producers and consumers is driven in a declarative manner: the Orchestration system interacts with DeviceRegistry, SystemRegistry, ServiceRegistry and PlantDescription systems to produce orchestration rules to individual application systems. Such orchestration data have to consider the QoS requirements set by individual application systems. These QoS requirements will be considered as constraints on the matchmaking.

The QoSSetup service will acts as a support service to the Orchestration system. For every change in a local cloud, the resulting QoS has to be predicted. The changes cause the Orchestration system to compute alternative orchestrations which should be verified through the QoSSetup service. This will be repeated until a specific set of orchestrations appears to support the required QoS. Once the orchestration is settled the Orchestration system requests the QoSSetup service to perform the reservations necessary to grant the QoS. The Orchestration system also distributes the service end points to the systems involved.

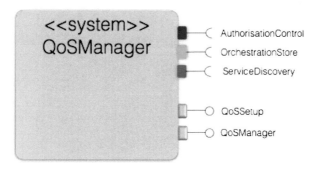

FIGURE 3.22
The QoSManager system and its produced and consumed services.

Thus, the QoSManager system's objective is to verify, manage, and guarantee QoS for services.

Apart from consuming the mandatory core services, the QoSManager system produces two services, QoSMonitor and QoSSetup, cf. Figure 3.22. In addition, the QoSManager consumes data from the SystemRegistry and the DeviceRegistry to deduce network topology and device capabilities within the local cloud. From this point of view, these registries provide information about the network tuning space. The current Arrowhead Framework view on QoS aspect and possible tuning spaces is further discussed in Section 4.3.5.

3.4.2.7 Historian system

The objective of the historian is to provide on demand the possibility to log service exchanges and store and retrieve any payload data produced by services registered within the local cloud. Thus the Historian system provides the possibility to store audit information as well as to keep historical record of data produced within a local cloud. Service data and audit information can be extracted using filters. Thus, enabling the extraction of, for example, audit data regarding a producer and its activity period, number of payloads provided, errors, etc.

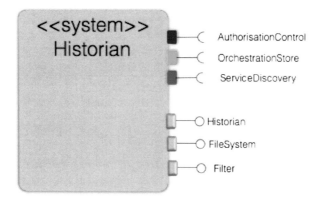

FIGURE 3.23
The Historian system and its produced and consumed services

The Historian should be able to store any service events created by any application service in a local cloud. Which services to store events from is the responsibility of the Orchestration system. Two types of application systems can be distinguished

- Application systems capable of directly consuming the Historian service

- Application systems making used of the EventHandler system to interact with the Historian system

  To extract data from the Historian is supported by two services:

- FileSys service

- Filter service

In summary, the Historian produces three services and consumes the mandatory core services, cf. Figure 3.23.

3.4.2.8   Gatekeeper system

Inter-cloud service exchange is essential to build real automation systems based on local clouds. Inter-cloud service exchange also supports scalability of a System of Systems.

For this purpose a local cloud need mechanisms that provide support for service discovery, system authentication and authorised service consumption, and data encryption. The scenario for establishing inter-cloud discovery, authentication & authorisation and orchestration visualised in Figure 3.25. Data encryption can still be maintained with e.g. IPSec or SOA protocol based encryption using for example MQTT; see the application discussion in Chapter 10.

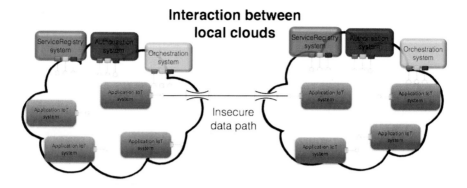

FIGURE 3.24

Service exchange between system residing in different local clouds may open an insecure data path. Such data a path punches a hole in the local cloud security wall, thus opening a path for intruders.

The Arrowhead Framework approach to this is called the Gatekeeper system. The Gatekeeper system objective is to enable the discovery, orchestration, and authentication and authorisation of services residing in different local clouds.

The Gatekeeper system produces two services and consumes the mandatory core services of its local cloud, cf. Figure 3.26. A detailed description of the Gatekeeper system and its services can be found in Section 4.3.7.

For the inter-cloud service exchanges it's also necessary to secure the data path hole in the local cloud security fence. For security reasons it's not desirable that two systems residing in different local clouds should be able to directly exchange services.

The Arrowhead Framework provides two approaches to a secure data path. One approach makes use of the well-known Demilitarized Zone (DMZ) [38]. The approach is based on a double historian approach which is described in detail in Section 3.4.2.9. The other approach, is based on the MQTT protocol and its broker technology; see a detailed description in Chapter 10 .

3.4.2.9 Historian-Historian secure data path

When an inter-cloud service exchange has been orchestrated and authenticated using the Gatekeeper system the next step is establishing of a data path between the local clouds.

Establishing the inter-cloud data path can be done in several ways:

- Direct between two services residing in two different local clouds. This will provide a security hole in the local cloud and thus,an opening for exposing the local cloud to external communication and, e.g., a denial of service attach.

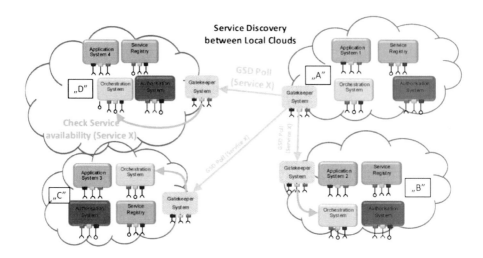

FIGURE 3.25
Inter-cloud service exchange supported by global dervice discovery, orchestration, and authorisation of service exchange.

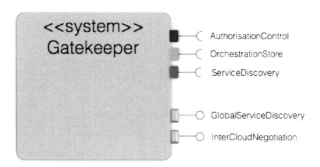

FIGURE 3.26
The Gatekeeper system and its produced and consumed services.

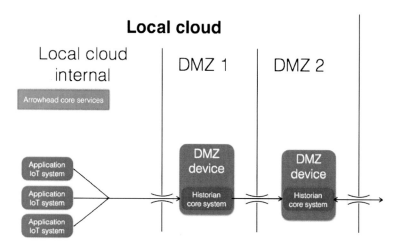

FIGURE 3.27

Establishing a secure inter-cloud data service patch by buffering data with a historian service which only can be accessed by another historian which in turn is exposed externally to the local cloud.

- To establish a secure data path Arrowhead Framework provides a type of demilitarized zone (DMZ) [38], solution. The buffering of application service data is made in the Historian system with an double DMZ layer approach as depicted in Figure3.27. The DMZ devices hosting the two Historian system have to have dual network interfaces thus, ensuring that no external traffic can reach the internal local cloud.

- Yet another way to establish the secure data path is by the use of the MQTT protocol and its broker technology; see Chapter 10 for details.

3.4.2.10   Translation system

The translation system is a transparency technology which resolves protocol, encoding, semantic, and security interoperability missmatches. An industrial IoT requires having multi-domain applications interacting seamlessly. Currently, industrial IoTs have many protocols which cannot communicate without specialised middleware or application layer multi-protocol interfaces. These solutions are both costly and not scalable. Hence an interoperability solution such as the translation system is required [39]. In Figure 3.28 a scenario with translation systems supporting interoperability between different services using incompatible SOA protocols is shown.

The objective of the Translation system is to provide translation of protocol, encoding, semantic and security between a service producer and a service consumer having non-interoperable SOA interfaces implementations.

The Translation system produces the Translation service and consumes

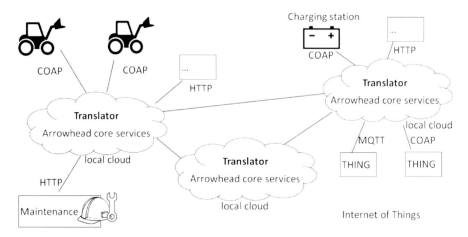

FIGURE 3.28

Service exchange scenario with translation systems supporting the interoperability between different services using different SOA protocols. Here protocol translation is supporting within a local cloud and in-between local clouds

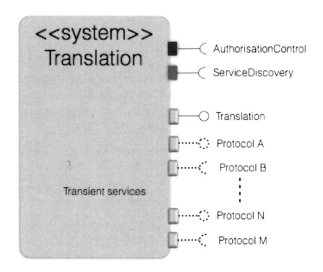

FIGURE 3.29

The Translation system and its produced and consumed services.

the mandatory core services, cf. Figure 3.29. The Translation service shall be consumed by the Orchestration system which, initiates the creation of a transient pair of consumed and produced service X. The consumed service X uses protocol A and the produced service X uses protocol B.

Interoperability between protocols can be achieved in different ways. In

Figure 3.30 there are three models of translator presented. In Figure 3.30-a the translation is made by direct translation. This model becomes very inefficient as the number of protocols increases. The number of required translators is given by Equation 3.1.

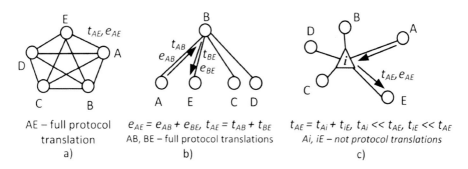

AE – full protocol translation
a)

$e_{AE} = e_{AB} + e_{BE}, t_{AE} = t_{AB} + t_{BE}$
AB, BE – full protocol translations
b)

$t_{AE} = t_{Ai} + t_{iE}, t_{Ai} \ll t_{AE}, t_{iE} \ll t_{AE}$
Ai, iE – not protocol translations
c)

FIGURE 3.30
Three different translation models possible in a protocol translator system.

$$\sum_{k=1}^{n-1} k = \frac{n(n-1)}{2} \qquad (3.1)$$

In Figure 3.30-b the translation is using an intermediate protocol, this is much more efficient in terms of translation implementations. However, it introduces latency and potentially additional loss of information. In 3.30-c translation uses an intermediate format, not constrained by the requirements of on-the-wire protocols, the translator is able to scale well and also maintain very good latency performance and information preservation.

Thus the Arrowhead Framework currently has defined a translation system according to each of the above given core systems; their services and interactions are described in detail in Chapter 4.

3.4.3 Application systems

Within the Arrowhead Framework the application system objectives are to implement application functionalities and services aiming to fullfil application requirements.

An application system may produce or consume other Arrowhead Framework services.

3.4.3.1 Application services

An application system is at minimum consuming the following mandatory core services; cf. Figure 3.31:

- ServiceDiscovery

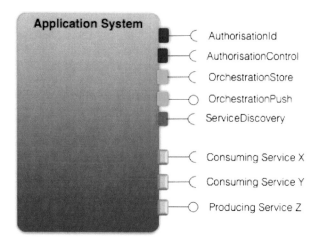

FIGURE 3.31
An application system is capable of consuming the Arrowhead Framework mandatory core services and will produce and/or consume one or more application services

- AuthorisationControl

- OrchestrationStore

In addition it is producing or consuming at leat one service. The application service may be capable of handling a service publish and subscribe schema. This shall be provided as metadata when registering the service with the ServiceRegistry.

Regarding use of authorisation the application shall provide meta-data if it can interact with an AA or an AAA authorisation system.

The application system may be capable of using the Historian system for data logging and audit information. This can be accomplished by an orchestration rule to the Historian system to consume the produced application service. For system that are resource constrained and have sleeping functionalities, it might be feasible for the application system to have the capability to consume the Historian service from the Historian system. Thus minimising administration of awake time slots and the uncertainties that missed time slots may create.

3.4.4   Deployment procedure for a local cloud

The Arrowhead Framework has architectural components that address the initiation of a local cloud and the invocation of trusted devices.

The objective of the deployment procedure is to create a local cloud in such a way that its basic functionalities and security have been established in a secure way. It is obvious that the deployment shall ensure that the mandatory core systems, their services and executing devices can be guaranteed to not be compromised.

Regarding enrolment of devices, systems and services into a local cloud there are bootstrapping mechanisms defined below. Of course a non-secure approach can be used here and just allow any device and its systems to register service into a local cloud. In most cases it is good practise to have a process that authenticates and authorises the entry of devices, systems, and services into a local cloud.

3.4.4.1 Secure bootstrapping of devices and systems into a local cloud

Assure that the ServiceRegistry system, the DeviceRegistry system, the SystemRegistry system, the Authorisation system, and the Orchestration system are established in a network in a secure way. Have these, non-compromised, systems executing on some devices. Thus the mandatory systems of a local cloud are established. To build a secure automation application the bootstrapping of application devices and systems, plus eventual support core systems and related devices, is critical from a security point of view.

To assure that the cloud is not compromised upon the introduction of systems, it is important to establish a chain of trust from service to system to device. For this purpose a secured initiation or bootstrapping process ranging from a device to its software system and associated services is needed. For the Arrowhead Framework, the proposed approach is based on a two-way clearance procedure of a device and its hosted software systems. The approach is based on a two way authentication.

To enable a device to be trusted, it has to have specific hardware providing storage and computation of authentication keys which shall be secure and tamper free. Such hardware, security controllers, are provided by a couple of vendors. The device further needs both a network interface and some type of short range communication, e.g., NFC. This will enable an operator identification of a device via key authentication over the NFC link. Such authentication will allow the generation of a device authentication key to be transferred to and stored in the security controller. The same procedure is to be made for each Arrowhead Framework system and each service produced by the system service. Thus the security controller holds the chain of trust including device key, system key and service key.

An Arrowhead Framework compliant system can then request to join a local cloud by requesting to

- register with the DeviceRegistry providing its device credential

- register with the SystemRegistry providing its system credential

These credentials, key plus ID, will now be used for identification with

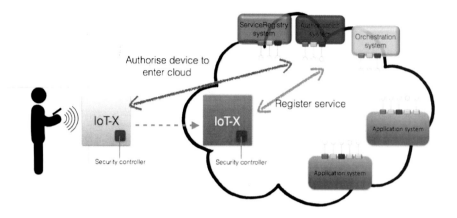

FIGURE 3.32

The bootstrapping process for a device, IoT-X, with IoT system A requesting to be entered into an Arrowhead Framework local cloud. A two-way authentication process is defined, involving human authorisation of the device and its hosted systems. Upon succesfull identification, the security controller in the device is provided with the keys allowing the Authorisation system to identify and admit the device and its systems into the local cloud.

the Authorisation system within the local cloud. First the device key is authenticated upon which the local cloud membership is authorised by the Authorisation system. The Authorisation system in responce provides the local cloud membership key to the device. Next each system hosted by the device is requesting membership of the local cloud using the same schema. This process is then repeated for each device and system to be registered with the local cloud. Finally all services produced by the registered systems shall be authenticated and registered with the ServiceRegistry system. This process of becoming a trusted device and system/s with trusted service/s within a local cloud is depicted in Figure 3.32. A sequence diagram for authenticated registration of a service with the ServiceRegistry is given in Figure 3.33. The sequence diagrams for a device and a system registering with their Registry are identical. In this way any device, system and associated service will be authenticated with a local cloud and a chain of trust is established.

The operator authentication approach should also be applied to the bootstrapping of the three mandatory systems plus the SystemRegistry and DeviceRegistry systems. Thus, providing a first level trust to these very core parts of a local cloud. Here the security controller within the device/s hosting these five systems and their services will store the operator interaction generated key together with the respective IDs.

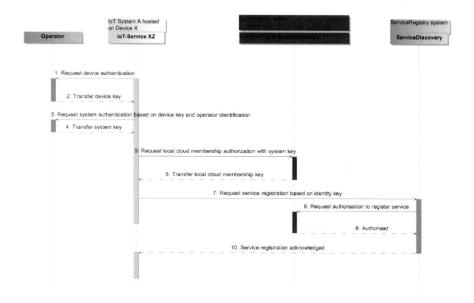

FIGURE 3.33

The sequence diagram for the process of authenticating a service (or system or device) with its registry thus allowing the registration of the service (or system or device) with its registry.

3.4.5 Creating Arrowhead Framework compliant systems

When designing an application system to operate within the Arrowhead Framework architecture devices, systems and services have to comply to the architecture principles. Following certain design steps will help to design an Arrowhead Framework compliant system and associated services. The following steps shall be performed to ensure compliancy to the Arrowhead Framework:

- Design according to Arrowhead Framework templates;

- Adapt legacy systems or implement new systems according to the Arrowhead Framework principles/patterns;

- Perform interoperability tests.

Besides the available design templates, the work of system architects and developers is supported through the design documentation guidelines defined in [19].

System design is often created by using already available legacy system components. In order to support interoperability for legacy systems, these should be adapted to the Arrowhead Framework. This can be done through adaptor or gateway components, which can be either integrated to the legacy systems (i.e., within the same hardware element), or can be made available as an Arrowhead core system with such gateway/adaptor capabilities.

To support interoperability testing, Arrowhead Framework provides a test tool which is further described in Section 5.6.2.

3.4.6 Interfacing to legacy systems

Due to the large amount of automation technology already installed there is a clear need for migration strategy and technology allowing the integration of the local cloud approach to legacy systems. The Arrowhead Framework automation architecture defines three levels of maturity. The maturity level indicates in what way an application system achieves conformance with the Arrowhead Framework; cf Figure 3.34. Three levels are defined:

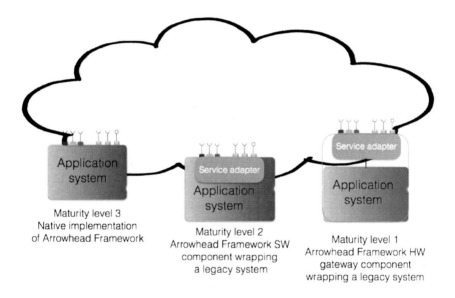

FIGURE 3.34
Application systems must publish information about their available services.

- **Level-3** The application system implements the consumption (AND/OR) production of Arrowhead Framework compliant services without using any external components.

- **Level-2** The application system implements the consumption (AND/OR) production of Arrowhead Framework services by using a software adaptor. The application system is thus modified by integrating the software adaptor into the system.

- **Level-1** The application remains unchanged. It uses dedicated hardware with software responsible for wrapping the application system with Arrowhead Framework compliant services. This hardware/software system implements a specific interface that can be connected to an existing system interface and proxies this information and functionality to Arrowhead Framework compliant services. The existing application system can remain unchanged.

3.4.7 Verification of Arrowhead Compliance

The purpose of the verification is to make sure that the system in question is compliant to the Arrowhead framework. The following is checked during the verification procedure.

- Can it connect and communicate properly with the mandatory core services?

- Does it comply with the rules for system documentation set for Arrowhead Framework compliant systems?

- Does it produce and consume services of the Arrowhead Framework as it is documented within its System Description (SysD)?

In order to technically validate the compliance, the Arrowhead Verification Tool has been created. It supports the following:

- System test and integration procedures through manual, automatic, and scripttests in order to verify and validate service realisation;

- Development through manual orchestration — to simplify producer and/or consumer interaction (with functionalities such as recording and playback of service interactions);

- Dynamic simulation of services and the handling of function chains, as well as service relations and their information exchange.

The compliance verification is to be controlled by personnell to evaluate operational as well as documentation aspects within different scenarios.

In conclusion, the Arrowhead Framework consists of what is needed for anyone to design, implement and deploy an Arrowhead compliant system aiming at enabling all of its users to work in a common and unified approach âĂŞ– leading towards high levels of interoperability, supporting the general objective of enabling information exchange between two IoT devices at a service level. Also further supporting the objective of enabling System of Systems operation. In turn enabling design, engineering and operation of collaborative automation systems using a local cloud approach addressing key properties of real time, security, engineering, and scalability.

For latest update on the Arrowhead Framework and the local automation cloud architecture, please consult the Arrowhead Framework wiki [25].

Bibliography

[1] A. W. Colombo, T. Bangemann, S. Karnouskos, J. Delsing, P. Stluka, R. Harrison, F. Jammes, and J. L. Lastra, "Industrial cloud-based cyber-physical systems," *The IMC-AESOP Approach*, 2014.

[2] S. Karnouskos, A. W. Colombo, F. Jammes, J. Delsing, and T. Bangemann, "Towards an architecture for service-oriented process monitoring and control," in *IECON 2010-36th Annual Conference on IEEE Industrial Electronics Society*. IEEE, 2010, pp. 1385–1391.

[3] T. Erl, *SOA Principles of Service Design (The Prentice Hall Service-Oriented Computing Series from Thomas Erl)*. Upper Saddle River, NJ, USA: Prentice Hall PTR, 2007.

[4] C. Pautasso, E. Wilde, and R. Alarcon, Eds., *REST: Advanced Research Topics and Practical Applications*. Springer, 2014.

[5] Z. Shelby, "The constrained application protocol (coap) - rfc 7252," IETF, Tech. Rep., 2014.

[6] "Xmpp is the open standard for messaging and presence," 2016. [Online]. Available: http://xmpp.org

[7] (2016) Mqtt is a machine-to-machine (m2m)/"internet of things" connectivity protocol. [Online]. Available: http://mqtt.org

[8] A. Banks and R. Gupta. (2016) Mqtt version 3.1.1. oasis standard. http://docs.oasis-open.org/mqtt/mqtt/v3.1.1/mqtt-v3.1.1.html.

[9] W. Mahnke, S.-H. Leitner, and M. Damm, *OPC Unified Architecture*. Springer, 2009.

[10] S. Cheshire and M. Krochmal, "Dns-based service discovery," IETF, Tech. Rep., 2013.

[11] P. Mockapetris, "Domain names - implementation and specification," IETF, Tech. Rep., 1987.

[12] R. Elz and R. Bush, "Clarifications to the dns specification," IETF, Tech. Rep., 1997.

[13] M. Cotton, L. Eggert, J. Touch, M. Westerlund, and S. Cheshire, "Internet assigned numbers authority (iana) procedures for the management of the service name and transport protocol port number registry," IETF, Tech. Rep., 2011.

[14] Wikipedia, "User datagram protocol — wikipedia, the free encyclopedia," 2016, accessed 20-April-2016. [Online]. Available: https://en.wikipedia. org/w/index.php?title=User_Datagram_Protocol&oldid=715583050

[15] J. Postel, "Rfc 768 user datagram protocol," IETF, Tech. Rep., 1980.

[16] "Extensible markup language - xml," 2016. [Online]. Available: https://en.wikipedia.org/wiki/XML

[17] D. Peintner and S. Pericas-Geertsen, "Efficient xml interchange (exi) primer," W3C, Tech. Rep., 2014.

[18] C. Jennings and Z. Shelby, "Media types for sensor markup language (senml) draft-jennings-senml-10," IETF, Tech. Rep., 2013.

[19] F. Blomstedt, L. L. Ferreira, M. Klisics, C. Chrysoulas, I. M. de Soria, B. Morin, A. Zabasta, J. Eliasson, M. Johansson, and P. Varga, "The arrowhead approach for soa application development and documentation," in *Proceedings IECON 2014*, 2014.

[20] "Arrowhead project." [Online]. Available: http://www.arrowhead.eu

[21] Wikipedia, "Unified modeling language — wikipedia, the free encyclopedia," 2016, [Online; accessed 19-April-2016]. [Online]. Available: https://en.wikipedia.org/w/index.php?title=Unified_Modeling_Language&oldid=708110201

[22] ——, "Business process model and notation — wikipedia, the free encyclopedia," 2016, [Online; accessed 19-April-2016]. [Online]. Available: https://en.wikipedia.org/w/index.php?title=Business_Process_Model_and_Notation&oldid=715814821

[23] ——, "Systems modeling language — wikipedia, the free encyclopedia," 2016, [Online; accessed 19-April-2016]. [Online]. Available: https://en.wikipedia.org/w/index.php?title=Systems_Modeling_Language&oldid=715050337

[24] ——, "Automationml — wikipedia, the free encyclopedia," 2015, [Online; accessed 19-April-2016]. [Online]. Available: https://en.wikipedia.org/w/index.php?title=AutomationML&oldid=673745347

[25] (2016) Arrowhead framework wiki. [Online]. Available: https://forge.soa4d.org/plugins/mediawiki/wiki/arrowhead-f/index.php/Main_Page

[26] E. Christensen, F. Curbera, G. Meredith, and S. Weerawarana, "Web services description language (wsdl) 1.1," W3C, Tech. Rep., 2001.

[27] M. Hadley, "Web application description language," W3C, Tech. Rep., 2009.

[28] Wikipedia, "Hateoas — wikipedia, the free encyclopedia," 2016, [Online; accessed 2-June-2016]. [Online]. Available: https://en.wikipedia.org/w/index.php?title=HATEOAS&oldid=715478952

[29] M. Myers, R. Ankney, A. Malpani, S. Galperin, and C. Adams, "X. 509 internet public key infrastructure online certificate status protocol-ocsp," RFC 2560, Tech. Rep., 1999.

[30] A. DeKok and A. Lior, "Remote authentication dial in user service (radius) protocol extensions," RFC 6929, April, Tech. Rep., 2013.

[31] K. Nagorny, R. Harrison, A. W. Colombo, and G. Kreutz, "A formal engineering approach for control and monitoring systems in a service-oriented environment," in *11th IEEE International Conference on Industrial Informatics (INDIN)*, July 2013, pp. 480–487.

[32] "Iec81364 - industrial systems, installations and equipment and industrial products – structuring principles and reference designations," ISO/IEC, Standard, 2009.

[33] O. Carlsson, D. Vera, J. Delsing, B. Ahmad, and R. Harrison, "Plant descriptions for engineering tool interoperability," in *Proceedings of IEEE INDIN 2016*, 2016.

[34] O. Carlsson, C. Hegedűs, J. Delsing, and P. Varga, "Organizing iot systems-of-systems from standardized engineering data," in *Proceeding IECON 2016*, Firenze, Italy, Oct. 2016.

[35] M. Albano, L. Ferreira, and J. Sousa, "Extending publish/subscribe mechanisms to soa applications." in *Proceedings of the 12th IEEE World Conference on Factory Communication Systems (WFCS)*, Aveiro, Portugal, May 2016.

[36] M. Albano, R. Garibay-Martínez, and L. L. Ferreira, "Architecture to support quality of service in arrowhead systems," in *Proceedings of IN-FORUM 2015*, Covilhã, Portugal, Sep. 2015.

[37] L. L. Ferreira, M. Albano, and J. Delsing, "Qos-as-a-service in the local cloud," in *Proceedings of SOCNE 2016, in conjunction with ETFA 2016*, Berlin, Germany, Sep. 2016.

[38] S. Jacobs, *Engineering Information Security: The Application of Systems Engineering Concepts to Achieve Information Assurance*. John Wiley & Sons, 2015, no. ISBN 9781119101604.

[39] H. Derhamy, J. Eliasson, J. Delsing, P. Varga, and P. Punal, "Translation error handling for multi-protocol soa systems," in *Proceedings of 2015 IEEE 20th International Conference on Emerging Technologies & Factory Automation (ETFA 2015)*, Luxembourg, Sept. 2015.

4

Arrowhead Framework core systems and services

Jerker Delsing
Luleå University of Technology

Jens Eliasson
Luleå University of Technology

Michele Albano
ISEP, Polytechnic Institute of Porto

Pal Varga
AITIA Inc

Luis Ferreira
ISEP, Polytechnic Institute of Porto

Hasan Derhamy
Luleå University of Technology

Csaba Hegedűs
AITIA Inc

Pablo Puñal Pereira
Luleå University of Technology

Oscar Carlsson
Midroc Automation

CONTENTS

4.1 Introduction

In Chapter 2 local clouds were discussed followed by a local cloud automation architecture in Chapter 3. The automation architecture supports the implementation of local automation clouds. Such implementation is supported by the Arrowhead Framework and its core systems and services.

The Arrowhead Framework core systems enable the creation and operation of local clouds. First implementation of these systems and their services are described in detail in this chapter.

There are currently two types of core services within the Arrowhead Framework:

- Mandatory core systems
  needed to establish the minimal local cloud.

- Automation support core systems
  extending local cloud capabilities intending to provide support for the design and operation of local automation clouds and interaction between local clouds.

4.2   Mandatory core systems and services

Mandatory core systems provide the minimum advisable services to establish a local automation cloud. The mandatory core systems are

- ServiceDiscovery system — Responsible for registering and enabling discovery of registered services.

- Authorisation system — Responsible for providing credentials to systems in the local cloud enabling system authentication and service exchange authorisation.

- Orchestration system — Responsible for providing service consumption patterns information to the system registered in the local cloud.

These mandatory core services will support the fundamental architectural properties assigned to a local cloud, as discussed in Chapter 3.

The Arrowhead Framework implementation of the mandatory core systems and their services provides these fundamental properties to a local cloud. Thus, enabling and allowing service exchanges between a service producer and a service consumer with the desired level of security and autonomy, which is briefly illustrated in Figure 4.1.

4.2.1   ServiceRegistry system

The Arrowhead Framework provides an implementation of the ServiceRegistry system. Implementing one of the three mandatory architectural systems to enable a local automation cloud. The ServiceRegistry system and its ServiceDiscovery service enable publication, lookup, and deletion of services in the service registry. The underlaying technology is DNS with the DNS-SD extension [1][2][3][4][5]. This adheres to well-known and proven Internet standards and technology.

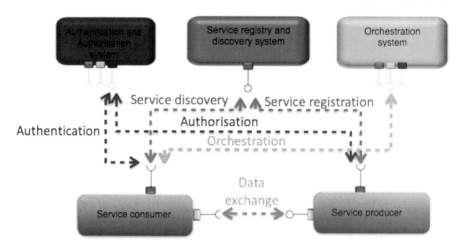

FIGURE 4.1
Minimal local cloud with indications of mandatory core service interaction enabling service exchange between two application systems.

The ServiceRegistry system is capable of storing reference information to all active producing services within the local cloud where it is active.

All systems within the local cloud that have services that produce information shall publish their produced service with the ServiceRegistry system by using the ServiceDiscovery service. A registration time out is recommended. For systems having sleep periods such meta data have to be provided. The DNS TXT record provide the location for time out and meta-data. The ServiceRegistry system uses this information to check for service availability in the local cloud. No response may result in service de-registration.

4.2.1.1   ServiceDiscovery service

The ServiceDiscovery service and its interface is defined according to Figure 4.2. The service interface provides three methods

- Publish
  The publish method is used to register services. The services will contain a symbolic name as well as a physical endpoint. The instance parameter represents the endpoint information that should be registered.

- Un-publish
  The un-publish method is used to unregister a service that no longer should be used. The instance parameter contains information necessary to find the service to be removed.

- Lookup
  The lookup method is used to find and translate a symbolic service name

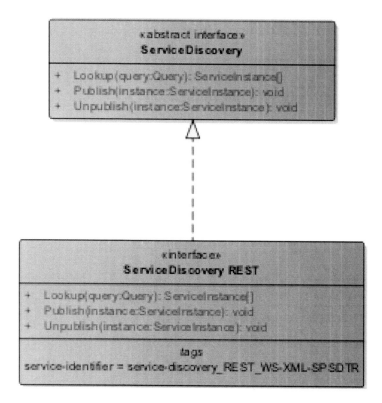

FIGURE 4.2
ServiceRegistry system produces the ServiceDiscovery service with the here depicted interfaces.

into a physical endpoint, IP address, and a port. The query parameter is used to request a subset of all the registered services fulfilling the demand of the requesting system. The returned listing contains service endpoints that fulfils the query.

The lookup, publish, and un-publish method sequences are provided in Figures 4.3, 4.4, and 4.5.

The information model follows the definitions in Section 3.2.6 and holds two data types:

- ServiceRecord
  with the following data:

  - Endpoint - string
    This datatype implements a representation of an endpoint using DNS A-records:

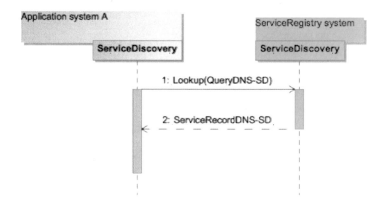

FIGURE 4.3
Sequence diagram for the lookup method of the ServiceDiscovery service.

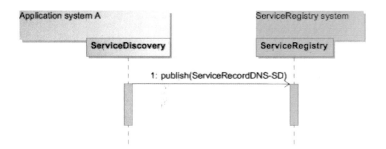

FIGURE 4.4
Sequence diagram for the publish method of the ServiceDiscovery service.

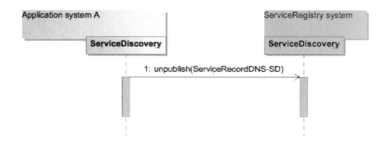

FIGURE 4.5
Sequence diagram for the un-publish method of the ServiceDiscovery service.

* Hostname is a String containing the name of the host in format: name.domain.topdomain, e.g., *app.arrowhead.eu*
* path: 192.168.1.20
* Port is an Integer containing the port number, e.g., 8070.
* Edata is a String containing additional information related to the endpoint. Any additional information that is required to identify the service instance should be stored in the mandatory DNS TXT record, as discussed in Section 3.2.6 and defined in [3].
* Metadata: key=value
  Metadata for the service are stored as key value pairs, could be, e.g., time to live, sleep period, configurations, payload encoding, compression and semantics. To allow for the orchestration to understand if any translation is necessary, it is proposed that the following three be mandatory:
  · Encoding: e.g., *encode=xml* where XML [6] encoding is used and specified in the CP (Communication Profile) document
  · Compression: e.g., *comp=exi* when EXI [7] compression is used and specified in the CP document
  · Semantics: e.g., *sem=senml* where SenML [8] semantics is used and specified in the SP (Semantic Profile) document
* ServiceName - string
  Name of the service instance e.g. *_Temp1*
* ServiceType - string
  e.g. *_ahf-temperature._coap._udp.*

- Query
  with the following content:

  – Query — string
    where the query string specifies one or several of the data types of the service. The query will then return a list of all registered services with the specified data type/s.

To access the ServiceRegistry from a REST based system, Arrowhead Framework has implemented the ServiceRegistryBridge system allowing easy ServiceRegistry interaction from a REST — http protocol based system.

The sequence diagram for REST based system interactions with the ServiceRegistry is shown in Figure 4.6.

The details of SLA meta-data and priority information in the service registry are beyond the scope of this book. Please refere to the Arrowhead Framework wiki for the latest details.

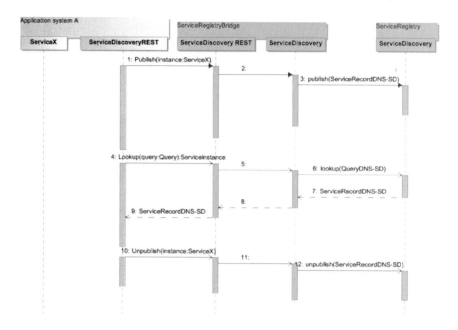

FIGURE 4.6

Sequence diagram a REST based system A registration process with the ServiceRegistry as supported by the ServiceRegistryBridge.

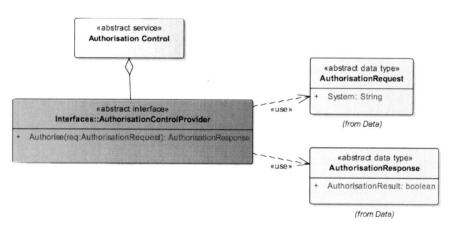

FIGURE 4.7

The AuthorisationControl interface with available methods and datatypes.

4.2.2 Authorization system

The Arrowhead Framework implements two different Authorisation systems. One is a authorisation and authentication, AA, system. The other is a authorisation, authentication and accounting, AAA, system. Both the AA and the AAA system meet the basic objectives requested by the Arrowhead Framework local cloud architecture.

4.2.2.1 AA-Authorisation system

Here an authorisation and authentication, AA, system is discussed. The AA-Authorisation system produces the AuthorisationControl service and the AuthorisationManagement service.

For the AuthorisationControl service the available interface is AuthoriseControlProvider; see Figure 4.7.

The Authorise Control Provider interface provides a clearance for a specific system to consume a specific service. The method is

- Authorise
  with the data types:

  - AuthorisationRequest
    requests the authorisation for the system addressed by an endpoint string e.g. `coap://192.168.2.30:8000/_viib3._ahf-vibration._udp`. The end point as provided by the orchestration system.

  - AuthorisationResponse
    responds with a boolean (True/False) request.

A sequence diagram for an application system A requesting authorisation to consume a specific method of a produced service is provided in Figure 4.8.

For the AuthorisationManagement service the available interface is AuthoriseManagementProvider; see Figure 4.9.

The AuthoriseManagementProvider interface provides a number of methods to manage authorisation rules:

- AddAuthorisationRules(rules:AuthorisationRulesList)
  is used to store new authorisation rules in the providing system.

- ListAuthorisationRules(): AuthorisationRules
  is used to present all the rules to the administrator of the authorisation ruling.

- ListAvailableServiceInstances(type:String): StringList
  is used to fetch all service instances that currently are stored in the authorisation system, in order to list consumer systems and producer systems. The purpose of this method is to allow an administrator to administer all systems.

- ListAvailableServiceTypes(): StringList
  is used to fetch the service tyes currently used in the arrowhead system-of-system, from the authorisation point of view.

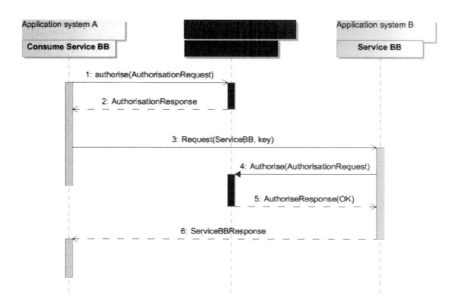

FIGURE 4.8

The sequence diagram for system A requesting authorisation to consume the interface definition and available data types of the AuthorisationManagement service.

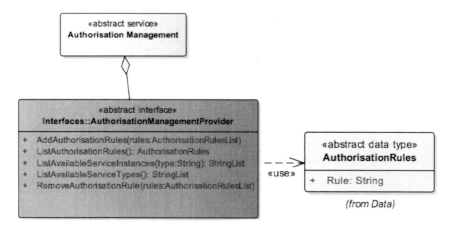

FIGURE 4.9

The AuthorisationManagement interface with available methods and data types.

- RemoveAuthorisationRule(rules:AuthorisationRulesList) is used to remove a rule that no longer is valid.

with the datatype:

- AuthorisationRules
  The AuthorisationRules data type contains information about rules that
  the consumer should be matched against in order to determine if a producer should be releasing information.

Sequence diagrams for the methods listAuthorisationRules and addAuthorisationRule are provided in Figure 4.10 and 4.11. Sequence diagrams for the
other methods can easily be deduced based on these sequence diagrams.

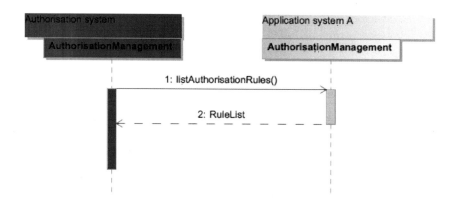

FIGURE 4.10
Sequence diagram for the list AuthorisationRule method of the AuthorisationManagement service providing a list of AuthorisationRules applicable to
a system A.

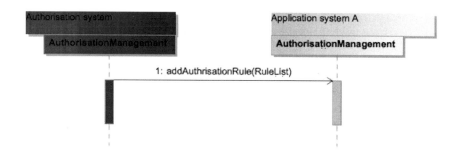

FIGURE 4.11
Sequence diagram for the addAuthorisationRule method of the Authorisation-
Management service providing AuthorisationRules to a system A.

The most recent information on the AA-authorisation system is available in the Arrowhead Framework wiki [9].

4.2.2.2   AAA-Authorization system

Here an authorisation, authentication, and accounting, AAA system is discussed. The Arrowhead Framework implementation of the AAA-Authorisation system and is based on Radius ticket technology.

The published services are

- AuthenticationID service
  providing the possibility to complete a Challenge-Response communication with the Authentication, Authorization and Accounting Server to get a new and an unique Ticket.

- AuthorisationControl service
  providing the possibility of enabling fine grained access control to any resource/service for external requests; also provides customised information about the external consumer.

- AuthorisationManagement service
  providing the possibility to manage the access control policies, accounts, accounting parameters, timeouts, etc.

The available interface of AuthenticationID is AuthenticationIDProvider, see Figure 4.12.

The AuthenticationIDProvider interface provides a resource for a new Ticket request based on a Challenge-Response. The method is

- Authenticate ID with the data types

  - AuthenticationRequest
    request a new and valid authenticator

  - ChallengeRequest
    returns the authenticator

  - ChallengeResponse
    sends the username and encoded password, based on the authenticator and SecretKey.

  - TicketResponse
    returns a valid ticket with the timeout.

The available interface of AuthorizationControl is AuthorizationControlProvider, see Figure 4.13.

The AuthorisationControlProvider interface provides a clearance for an specific system to consume an specific service. The method is

- Authorise with the data types

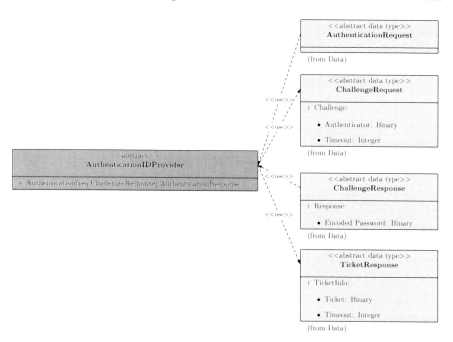

FIGURE 4.12
The interface definition and available data-types of the Authentication service.

- AuthorisationRequest
  request a validation of an specific ticket from an specific remote consumer.

- AuthorisationResponse
  returns the validity of a specific ticket, returning extra information about the owner of that ticket.

The available interface of the AuthorizationManagement service is AuthorizationManagementProvider; see Figure 4.14.

The AuthorizationManagementProvider interface provides a number of methods to manage authorization policies

- Authorise with the data types

  - ListPolicies():
    to list all available policies.

  - AddPolicy(policy):
    to add a specific policy.

  - RemovePolicy(policy):
    to remove a specific policy.

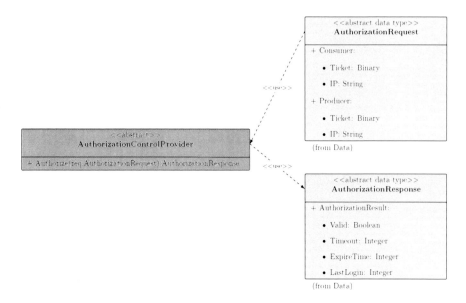

FIGURE 4.13
The interface definition and available data types of the AuthorisationControl service.

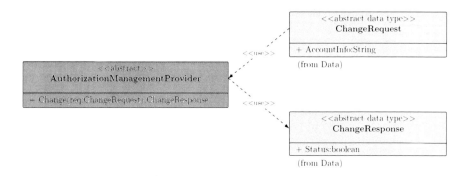

FIGURE 4.14
The interface definition and available data types of the AuthorisationManagement service.

- ModifyPolicy(policy,new_policy):
  to modify an specific policy.
- ListAccounts():
  to list all accounts.
- AddAccount(accountInfo):
  to create a new account.

- RemoveAccount(accountInfo):
  to remove a specific account.

- ModifyAccount(accountInfo,new_accountInfo):
  to modify an specific account.

The most recent information on the AAA-Authorisation system is available in the Arrowhead Framework wiki [9].

4.2.2.3   MMI-Authorisation system

Within the Arrowhead Framework an MMI-Authorisation system has been defined and implemented to support operator interaction with the Authorisation systems.

The MMI-Authorisation system provides a graphical user interface that allows a user/operator to manage and create access control rules for service producers.

The MMI-Authorisation system use the service AuthorisationManagement to communicate with the Authorisation system. The ServiceDiscovery service is used to list possible service for which access rules can be set.

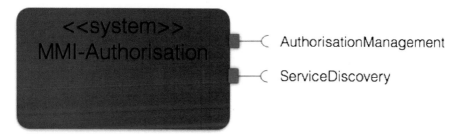

FIGURE 4.15
The MMI-Authorisation system and the services it shall consumed.

The MMI-Autorisation system do consume the AuthorisationManagement and ServiceDiscovery services, cf. Figure 4.15.

4.2.3   Orchestration system

The Arrowhead Framework implementation of the Orchestration system provides both the OrchestrationStore service and the Orchestration service.

The Orchestration system stores orchestration rules and resulting orchestration patterns. The requirements are specified during the design phase using engineering tools. These are provided to Orchestration system either via the PlantDescription service or directly from the engineering tools. New requirements can be provided in runtime upon which the Orchestration system computes new orchestration patterns which are provided to the involved application systems.

4.2.3.1   Orchestration services

The Arrowhead Framework implementation of the Orchestration system produces four services OrchestrationStore, OrchestrationPush, OrchestrationCapability and OrchestrationManagement. The Orchestration system also consumes services of type _ahfc-servprod_. There are currently two automation support core service of this type, the EventHandler service; see Section 4.3.4 and the Translation service; see Section 4.3.8.

OrchestrationStore service

The OrchestrationStore service provides functionality for storing and retrieving orchestration requirements. An orchestration requirement is a set of rules which describe the ideal service required by a consuming system. This could be as simple as a fully defined service contract. But it could also be as complex as describing service instance requirements, such as physical or geographic qualities.

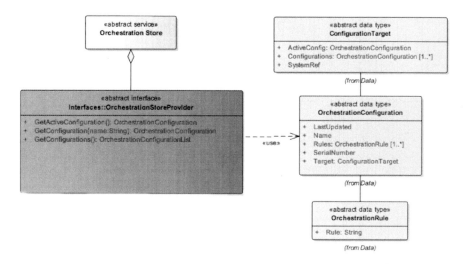

FIGURE 4.16
OrchestrationStore interface with methods and associated data definitions.

The OrchestrationStore interface with methods and associated data definitions is shown in Figure 4.16. The methods and associated data definitions are

- GetActiveConfiguration method is used to retrieve the currently active configuration. That means the configuration that should be executed instantly.

- GetConfiguration method is used to get a specific configuration based on the name of the configuration.

- GetConfigurations method is used to retrieve all configurations for any system, i.e., the orchestration configurations for the whole System of Systems.

- ConfigurationTarget data type defines a system that can be orchestrated.

- OrchestrationConfiguration data type defines how a system should be configured via a set of orchestration rules.

- OrchestrationRule datatype defines a connection between a producer and a consumer. OrchestrationRule contains end point information and consumption pattern enabling.

OrchestrationPush service

The Orchestration Push service is used to push orchestration configuration to an application system.

The consumer of the service has orchestration rules that describe a desired "end-state". The producer of the service is a system that has application services that should be connected to producers, in the Arrowhead network. By pushing the configurations from the consumer to the producer the receiving system will be able to connect its application service consumers.

The OrchestrationPush interface with methods and associated data definitions is shown in Figure 4.17. The methods and associated data definitions are

- PushOrchestration method is used to send orchestration configurations to the system that produces the service.

- ConfigurationTarget data type defines a system that can be orchestrated.

- OrchestrationConfiguration data type defines how a system should be configured via a set of orchestration rules.

- OrchestrationRule defines a connection between a producer and a consumer. OrchestrationRule contains end point information and consumption pattern.

The message sequences for PushOrchestration method is provided in Figure 4.18.

OrchestrationCapability service

Orchestration Capability service provides the capability to extract the current state in which service is provided by which system and which system is consuming which service enabling possibility to relate systems and their services. The OrchestrationCapability can thus be used to extract the current state of the enabled service exchanges. The state information should be extracted

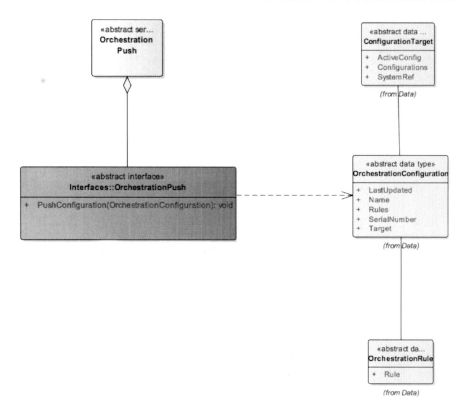

FIGURE 4.17
OrchestrationPush interface with methods and associated data definitions.

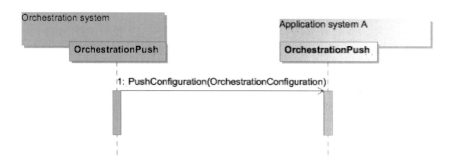

FIGURE 4.18
Sequence diagram for pushing orchestration configurations to an application system.

from the OrchestrationStore. This provided that all systems have released the orchestrated right to the Orchestration system.

Provided that such state detection can be triggered simultaneously across all local clouds at a production site, essential state information and topology information can be gathered through the OrchestrationCapacity service.

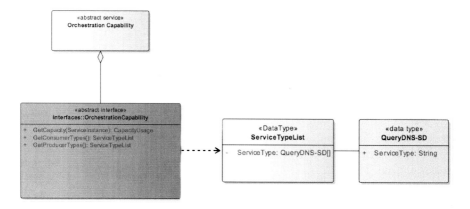

FIGURE 4.19
OrchestrationCapability interface with methods and associated data definitions.

The OrchestrationCapacity interface with methods and associated data definitions is shown in Figure 4.19. The methods and associated data definitions are:

- GetCapacity method returns a listing of available connections a service supports as well as the number currently connected to the service.

- GetConsumerTypes method returns a list of consumer types that the system will support.

- GetProducerTypes method returns a list of producer types that the system will be able to produce.

- QueryDNS-SD data type implements the Query data structure using DNS address lookup. ServiceType is a string identifying the service name.

OrchestrationManagement service

The OrchestrationManagement service handles management of orchestration configurations, including creation, editing, and removal of these configurations.

The OrchestrationManagement interface with methods and associated data definitions is shown in Figure 4.20. The methods and associated data definitions are

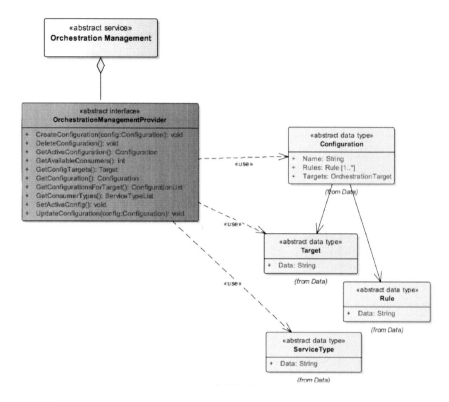

FIGURE 4.20

OrchestrationManagement interface with methods and associated data definitions.

- CreateConfiguration method is used to create orchestration configurations

- DeleteConfiguration method is used to remove configurations for a specific target

- GetActiveConfiguration method is used to retrieve the currently used orchestration configuration

- GetAvailableConsumers method is used to determine the number of consumers of a specific type the system can serve simultaneously

- GetConfigTargets method is used to get a listing of the systems that can be orchestrated

- GetConfiguration method is used to fetch current orchestration configuration for the specific system

- GetConfigurationsForTarget method is used to fetch all orchestration configurations for a specific target

- GetConsumerTypes method is used to list the supported service types a system can consume

- SetActiveConfig method sets which orchestration configurations that are to be used

- UpdateConfig method is used to update a specific orchestration configuration

- Configuration datatype defines how a system should be configured via a set of orchestration rules.

- Rule datatype defines a connection between a consumer and a producer

- ServiceType datatype defines the type of service that a system provides Ideally this data type should be retrieved from the service registry service

- Target datatype defines the orchestration target, that is a system that should be affected by the orchestration rule(s)

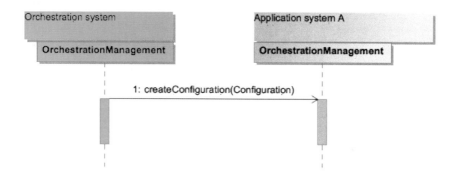

FIGURE 4.21
Sequence diagram for the dynamic behaviour when creating orchestration configurations.

The message sequences for all these interfaces is straight forward. Thus only the sequence diagram for CreateConfiguration is provided; cf. Figure 4.21.

4.2.3.2 MMI-Orchestration system

To manage the Orchestration system a GUI is provided through the optional MMI-Orchestration system. It provides a graphical user interface that allows an orchestration manager to create connection rules (orchestration) for systems (i.e., for system A and system B). The MMI-Orchestration system consumes two services, cf Figure4.22:

- ServiceDiscovery — to register itself into the local cloud

- OrchestrationManagement — where an orchestration manager can create or change an orchestration rule.

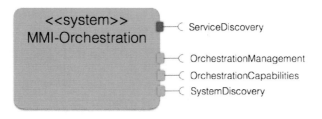

FIGURE 4.22
The MMI-Orchestration system and its consumed and produced services.

It is obvious that complex orchestration rules have to be generated by MES tools or similar. For now this is beyond the scope of this book.

4.3   Automation support core systems

To support the Arrowhead Framework architecture a number of automation support systems and associated services have been defined. Implementations of several of these are already available at the Arrowhead Framework wiki [9]. These systems and servies are described in detail below.

4.3.1   PlantDescription system

The purpose of the PlantDescription system is to provide a basic common understanding of the layout of a plant or site, providing possibilities for actors with different interests and viewpoints to access their view of the same dataset. These datasets are to be provided by the databases integrated in or generated by the flora of computer-aided design (CAD) tools used by engineers from all disciplines involved in designing a large automation System of Systems. More detailed engineering scenarios, including intended usage of the PlantDescription system, can be found in Section 6.2. It is clear that one plant description can span several local automation clouds.

A source of inspiration for the design was the standard ISO/IEC 81346 [10], which specifies that for studying objects and their relations it may be useful to look at them from different viewpoints, highlighting different aspects of the objects and relations. This standard is focused on the three aspects function, product, and location, although the design is intended to be capable of addressing other viewpoints as well.

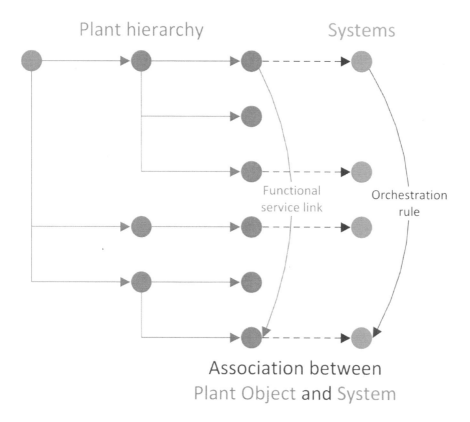

FIGURE 4.23
Distinctions between system responsibilities. Only the blue part of this diagram should typically be provided by the plant description.

In Figure 4.23 different types of information about a plant are illustrated.

- Blue represents the fairly static structure of a plant, such as locations, mechanical hardware, process equipment, or expected overall functionality.

  - This is what is typically provided by the plant description.

- Green represents the Arrowhead systems that are operating within the plant.

  - Information about the systems is typically stored as metadata within the systems themselvs but also in the ServiceRegistry, the System-Registry, and the DeviceRegistry.

- Orange illustrates the link between the overall structure of the plant and the systems, and thereby shows how a system fits into the larger scale purpose.

- These links may be stored by the systems themselves, in a configuration system, and/or as metadata in the SystemRegistry. For replacement purposes it may be useful to have the information stored outside of the system itself.

- Purple represents an orchestration rule, used to direct Arrowhead Framework systems to consume services of other Arrowhead Framework systems.

  - Orchestration rules are generated by the PlantDescription system and provided to the Orchestration system.

Further details on the PlantDescription architectural and technology details can be found in [11, 12].

4.3.1.1   PlantDescription services

The two functions GetViewpoint and GetObjectsByNode are the main points of access to the PlantDescription system. An optional function GetObjectsByTypeId can be implemented to be used as a translator from one viewpoint to another.

GetViewpoint(Type): PlantData

The function GetViewpoint is used with an argument (Type) to get all nodes and links that correspond to this type.

GetObjectsByNode(NodeId): PlantData

The function GetObjectsByNode is used with a specific NodeId to get all nodes linked to the specified node, and those links.

GetObjectsByTypeId(Type, NodeTypeId): PlantData

The function GetObjectsByTypeId is used with a specific combination of a Type and a NodeTypeId corresponding to the identity by which the node is identified according to that type. The function returns the same information as the GetObjectsByNode for that node.

4.3.1.2   Abstract information model

The information provided by the Plant description service consists of sets of nodes and links that can be used to describe different topologies, hierarchies, and structures for a large facility. Figure 4.24 describes the attributes of the three data types PlantData, Node, and Link.

The PlantData, Node, and Link datatypes are provided in Tables 4.1 4.2 and 4.3. In Table 4.4 a set of service metadata is provided.

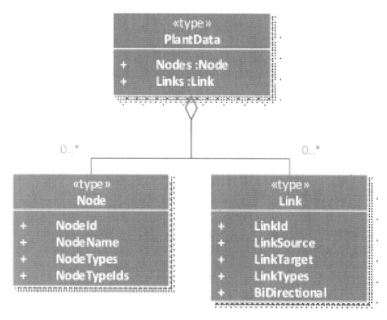

FIGURE 4.24

PlantData structure.

TABLE 4.1

PlantData data type description

Field	Description
Nodes	A collection of (zero or more) nodes
Links	A collection of (zero of more) links

TABLE 4.2

Node data type description

Field	Description
NodeID	A name or description for the node that is useful for human interaction (optional)
NodeTypes	A collection of (zero of more) node types
NodeTypeIds	An identifier or name specific for each type that the node belongs to (optional)

4.3.2 Configuration system

The purpose of the Configuration system is to provide a uniform way for Arrowhead compliant system to manage distribution of configurations. The extent of the configurability of a system ultimately depends on the system itself and therefore the design of this service is intended to allow different levels

TABLE 4.3

Link:Data data type description

Field	Description
LinkID	A unique identifier for the link
LinkSource	The NodeId of the source node
LinkTarget	The NodeId of the target node
LinkTypes	A collection of (zero or more) link types
BiDirectional	Parameter describing if the link is bi-directional

TABLE 4.4

Service metadata description

Field	Description	Mandatory
Plant	Identifies the plant, site, area, network, or similar that the instance contains information on	Yes
ViewpointTypes	List of Viewpoint types supported by the instance	No
LinkTarget	The NodeId of the target node	Yes

of configurations to be transferred using the same interface, from changes in system parameters to full firmware updates that may change which services a system is able to produce and consume and other fundamental properties.

Through this design the decision of how configurable a device or software system should be is left to the provider, and not imposed by the framework, while still allowing a uniform method for configuration management across diverse Systems of Systems containing devices and software systems with different levels of configurability.

The focus of the initial design of the Configuration system has been on the interface towards application systems, the service ConfigurationStore, indicating that further development of the service ConfigurationManagement may be desired once more partners take an interest in developing powerful management tools for the management of Arrowhead systems.

4.3.2.1   Configuration services

To provide the functionality included in this design the Configuration system provides two services, the ConfigurationStore and ConfigurationManagement. The ConfigurationManagement service is intended to be used by a user or other system to assign configuration files to configurable devices while the ConfigurationStore is the service to be consumed by systems on configurable devices to access the configuration files that have been assigned to them.

Depending on the which automation support core systems are available, what capabilities the configurable devices/systems have, and what security

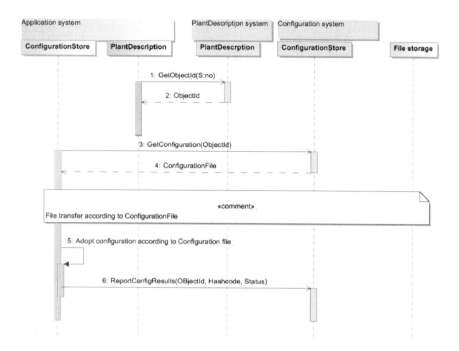

FIGURE 4.25
Sequence diagram for configuration of a device through the PlantDescription and Configuration systems.

and identification mechanisms are used, the interaction scheme might look slightly differently. Figure 4.25 illustrates how the interaction might look in a scenario where the PlantDescription system is used to track which physical device is used to implement a specific function in the larger System of Systems. Here the device identifies itself using its serial number, and there are no explicit authentication and authorization interactions included. In a different implementation the identity used for interaction with the ConfigurationStore service may be provided to the hardware device or hosted software system during an initial deployment process, either upon request or as part of an earlier registration with the Authorization system or an DeviceRegistry system.

A special usage of the ConfigurationStore service is as a deployment mechanism for software systems, as discussed in Section 3.4.4. The authorisation of the ConfigurationStore service to configure a device is provided by the local cloud Authorisation system. In this way ConfigurationStore is supporting bootstrapping of Arrowhead Framework compliant services to a trusted device within a local cloud.

4.3.3 SystemRegistry and DeviceRegistry systems

Both the SystemRegistry and DeviceRegistry systems are in most aspect, a carbon copy of the ServiceRegistry system. The main and only diference are that they shall register and store which systems and devices currently are registered with the local cloud.

Currently no implementations are available of these two systems. Such implementation should, however, be straightforward to implement based on the ServiceRegistry system and the usage of DNS and DNS-SD technologies.

The SystemRegistry system is capable of storing reference information to all active systems and their produced services and their hosting devices. Consequently, the DeviceRegistry system is capable of storing reference information to all active devices and their hosted systems.

4.3.3.1 SystemDiscovery and DeviceDiscovery services

The SystemDiscovery and DeviceDiscovery services interfaces are defined according to Figure 4.26.

The service interfaces provide three methods:

- Publish
  The publish method is used to register systems or devices. The services will contain a symbolic name as well as a physical endpoint. The instance parameter represents the endpoint information that should be registered.

- Unpublish
  The unpublish method is used to unregister a service that no longer should be used. The instance parameter contains information necessary to find the service to be removed.

- Lookup
  The lookup method is used to find and translate a symbolic service name into a physical endpoint, IP address, and a port. The query parameter is used to request a subset of all the registered services fulfilling the demand of the requesting system. The returned listing contains service endpoints that fulfil the query.

The lookup, publish, and unpublish method sequences are provide in Figures 4.27, 4.28, and 4.29. The sequence diagrams for the DeviceDiscovery methods are almost identical.

The information model for the publish method holds two data types:

- SystemRecord
  with the following data:

    - Endpoint - string
      This data type implements a representation of an endpoint using SRV record of DNS.

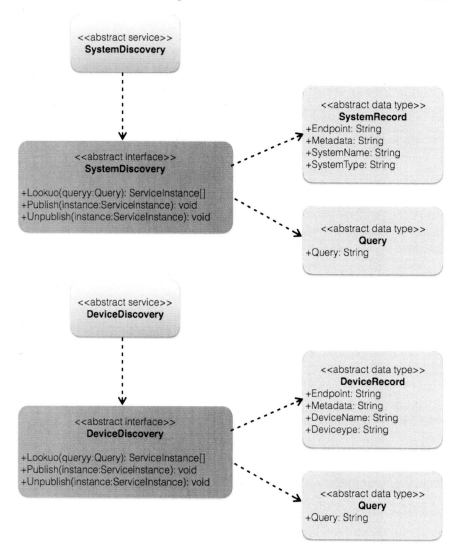

FIGURE 4.26
SystemDiscovery and DeviceDiscovery services with their three interfaces and associated SystemRecord and DeviceRecord data types and Query data type.

* Hostname is a string containing the name of the host in format: name.domain.topdomain, e.g., *app.arrowhead.eu*
* path: 192.168.1.20
* Port is an Integer containing the port number, e.g., 8070
* Edata is a String containing additional information related to the endpoint. Any additional information that is required to identify

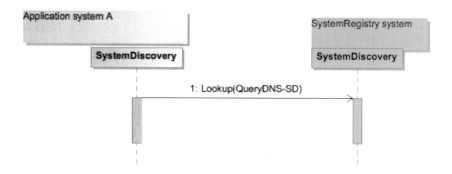

FIGURE 4.27
Sequence diagram for the lookup method of the SystemDiscovery service.

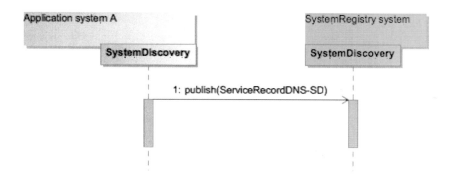

FIGURE 4.28
Sequence diagram for the publish method of the SystemDiscovery service.

the service instance should be stored in the mandatory DNS TXT record, as discussed in Section 3.2.6 and defined in [3]

* Metadata - string
  Metadata for the service are stored as key value pairs, e.g., time to live, sleep period, configurations, payload encoding, compression, and semantics. To allow for the orchestration to understand if any translation is necessary, it is proposed that the following three be mandatory:

  · _encoding is, e.g., _xml where XML [6] encoding is used and specified in the CP (Communication Profile) document
  · _compression is, e.g., _exi where EXI [7] compression is used and specified in the CP document

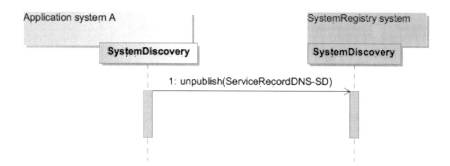

FIGURE 4.29
Sequence diagram for the unpublish method of the SystemDiscovery service.

> · _semantics is, e.g., _senml where SenML [8] semantics is
> used and specified in the SP (Semantic Profile) document
> * ServiceNamn - string
> Name of the service instance, e.g., _ Temp1.
> * ServiceType - string
> e.g., _ahf-temperature._coap._udp.

- Query
  with the following content:

  - Query - string
    where the query string specifies one or several of the data types of
    the service. The query will then return a list of all registered services
    with the specified data type/s.

To access the ServiceRegistry from a REST based system Arrowhead
Framework has implemented the ServiceRegistryBridge system allowing easy
ServiceRegistry interaction from a REST—http protocol based system.

The sequence diagram for REST based system interactions with the ServiceRegistry is shown in Figure 4.30.

4.3.4   EventHandler system

The EventHandler core system is in charge of the notification of events that
occur in an Arrowhead-compliant installation [13]. The Eventhandler system
is available to support scenarios like:

- Providing resource and functional constrained service producers with publish subscribe capability

- Reduce service consumption load on resource-constrained producers

the service producer are connected by a wide area network such as the Internet, the in-between network is not under control, real-time objectives are not feasible, and either prioritisation can be provided by means of solutions similar to Diffserv [16], or communication must operate in best effort.

The setup of QoS is a process strongly correlated with the service orchestration, performed through the Orchestration service. Arrowhead services are consumed after they are orchestrated together and obtained through interaction with the Orchestration service of the Orchestration system, and this latter system is in charge of accessing the QoSSetup service of the QoSManager system, which takes care of verifying that the QoS objectives are feasible, of keeping track of resource reservation, and of configuring the network actives and the devices according to the QoS objectives. Thus, this paradigm implies that the service request sent to the Orchestration Systems contains both the functional requirements of the service and the QoS objectives, which are expressed by means of a Service Level Agreement (SLA).

Regarding the monitoring of the QoS, the system offering the QoSMonitor service monitors the performance of services, either directly by having modules running over network actives and systems, or by accessing logs of network actives and systems, to detect if QoS parameters are not being guaranteed by the currently orchestrated service instance; in this latter case, the QoSMonitor service will contact either the Orchestrator system or the service consumer through the EventHandler system, to instruct them to repeat the orchestration process. Depending on the configuration at hand, the service can be offered by the QoSManager system itself, or by a system positioned strategically in the System of Systems. In the latter case, the QoSMonitor service will be used by the QoSManager system to configure its monitoring operations.

An example of exchange of messages related to the orchestration of a service offering QoS is given in Figure 4.32.

4.3.5.1   QoSSetup service

The QoSSetup service is used to verify the feasibility of QoS objectives for a given orchestration. The information returned by the QoSSetup service influences the Orchestrator system to change device and/or system network and computational parameters to find a setting for which the QoS constraints can be met. The response to the Orchestration system is either QoS-OK or QoS-not-OK. On receiving a QoS-not-OK response, the Orchestration system has to provide a new orchestration proposal. It is possible that the proposed procedure may end up in an endless number of iterations, but the problem of the convergence of the QoS and Orchestration interaction is beyond the scope of this book and a matter for further research.

The QoSManager shall be able to retrieve data from the ServiceRegistry, SystemRegistry, DeviceRegistry, OrchestrationStore, and ConfigurationStore within the local cloud. Based on data from these registries and stores, the QoSManager system will be able to deduce QoS constraints as well as available

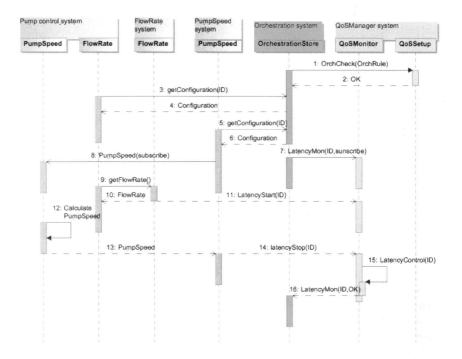

FIGURE 4.32
Sequence diagram for the orchestration of a simple control loop involving QoSSetup and QoSMonitor services.

computational and network resources. The algorithms used for computing QoS feasibility and for smart QoS management are beyond the scope of this book.

4.3.5.2 QoSMonitor service

The QoSMonitor service performs online monitoring of the performance of the services, network actives, and devices hosting the systems, aiming at verifying whether QoS objectives are attained or not.

A simple latency monitoring process is visualised in Figure 4.32, where the QoSMonitor is orchestrated to subscribe to the service exchanges being part of time critical paths. The services exchange time-stamped data, enabling the QoSMonitor to calculate total latency from source to receiver.

It is clear that more advanced monitoring of QoS may be necessary within the local clouds of larger automation systems. Anyway, this topic is beyond the scope of this book and a matter for further research.

4.3.6 Historian system

The Historian system is used for storage, processing, and visualisation of data created by services in a local cloud, e.g., sensor data. Since the Historian system should be able to consume a large number of simultaneous active services, it should be executed on a sufficiently fast computer.

Since a Historian system may interact with a potentially very large number of heterogeneous devices and systems, it is beneficial for it to support as wide range of protocols and data models as possible.

The Historian system can be orchestrated to consume any service in the local cloud. The consumed data from a specific service will be stored in a file being named as the consumed service endpoint. Data stored are service payload plus metadata provided by the used protocol.

4.3.6.1 Historian service

The produced Historian service has one interface: PutData, with the data types shown in Figure 4.33. Every access to the PutData interface will create a file in the Historian with a filename of (service_endpoint.timestamp).

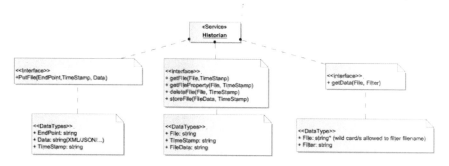

FIGURE 4.33
Historian interfaces and data types.

Figure 4.34 shows the procedure for a sensor platform that wants to store its sensor reading in a Historian system.

4.3.6.2 FileSys service

Service consumers, which can be human operators or other systems, that need to obtain data stored in the Historian system can use the FileSys service. The FileSys service has interfaces and data types shown in to Figure 4.35.

Figure 4.36 shows the procedure for an application system that will retrive and subsequently delete data from the Historian system.

This service enables clients to store and delete files and folders like a distributed file system. This service also supports generation of HTML data for creating web-based interfaces for all files and folders that are stored in a Histo-

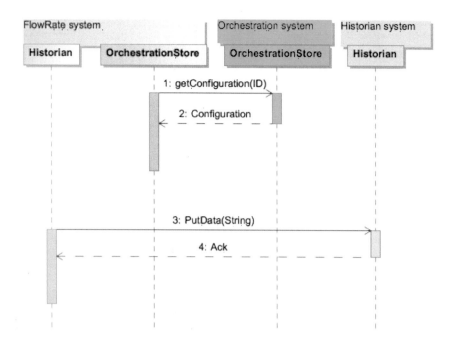

FIGURE 4.34

Application system interaction to store data in the Historian system data storage.

rian system. This allows sensor data, or any type of files, to be stored, viewed, and downloaded using standard web browsers. The HTTP-based FileSystem service also enables third party software, e.g., MATLAB, Simulink, or Excel, to be used for data processing and visualisation.

4.3.6.3 Filter service

The Filter service enables application consumers to retrieve data from the Historian based on the content of the stored data.

For the latest development of the Historian system and its service please refere to the Arrowhead Framework wiki.

4.3.6.4 Service information data

The Historian's data model is primarily based on the SenML specification. Encoding is either JSON, CBOR, or XML. Other formats may be supported and can be queried at run time. The Historian understands and decodes all units currently defined in SenML and can take appropriate actions depending

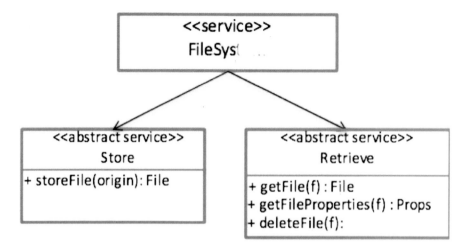

FIGURE 4.35
Interface definitions of the FileSys service.

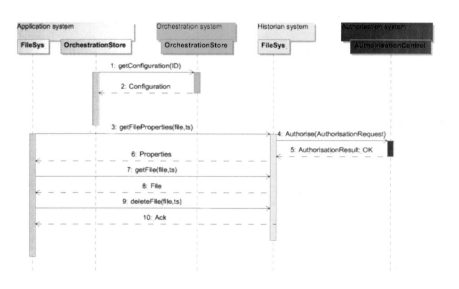

FIGURE 4.36
Sequence diagram for Historian system data retrieval and deletion.

on filter rules, etc. This service can also be used to monitor when new systems, and existing ones for that matter, initiate data communication.

4.3.6.5 Service metadata

The Historian also supports dynamic querying about active devices and systems. Clients can ask a Historian for all devices or systems, i.e., data sources, that have been active, for example, current day, last week, etc. Errors are represented using the SigML format. When using the file system feature, the Historian system is encoding and semantics agnostic.

The current implementation of the Historian's services supports a multitude of protocols, encodings, and data formats. For the latest specifications please consult the Arrowhead Framework wiki `https://forge.soa4d.org/plugins/mediawiki/wiki/arrowhead-f/index.php/Support_Core_Systems_and_Services#Historian_system`.

4.3.7 Gatekeeper system

There is a valid need for inter-cloud servicing, as one single Arrowhead cloud cannot serve all demands. Here are some examples (use cases) when inter-cloud relations are necessary or simply better than handling requests at a local level:

- Servicing is not possible: no service provider locally.

- Service provider is not available: is in out-of-order status or registered, but not found at the time.

- Servicing is currently not possible: no free resources or the QoS expectations cannot be met locally.

- Servicing within the home cloud is not optimal (e.g., geographically a different cloud is closer or better).

One of the major philosophical questions here is about data ownership, whether the command over servicing belongs to the system (e.g., it can choose its partners) or to the managing central entities of the local clouds (and these core systems pair up systems for servicing). As various intelligent decision-making processes (e.g., resource management) are implied here, in our primary scenario we assume that

- data ownership belongs to the local clouds (but this is not a necessity of the Gatekeeper services concept)

- one identity of an Arrowhead System of Systems cloud is requesting a service that cannot be fulfilled in the local cloud and therefore

- it might be viable to look for clients in other Arrowhead clouds

The Gatekeeper system provides the following functionality (see Figure 4.37):

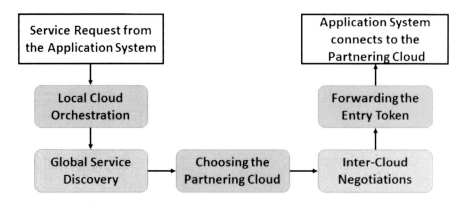

FIGURE 4.37
Creating inter-cloud consumer-producer relationship methodology.

1. *Global service discovery*: Finding other Arrowhead local clouds with suitable providers for the requested purposes.

2. *Inter-cloud negotiations*: authentication, identity verification of the chosen partnering cloud, authorising transactions, and establishing secure connections by managing the dialogs ("negotiations") between the clouds for establishing the inter-cloud producer-consumer relationship.

4.3.7.1 Global service discovery

There are several approaches to track the available services outside the local cloud. These approaches are suitable for different environments and requirements depending on the regularity of inter-cloud interactions, the volatility of service availability in the different clouds and the flexibility required from the collaborating Arrowhead System of Systems.

The most basic approach is hardwiring — and manually configuring — information about a certain set of other Gatekeepers into each Gatekeeper they can turn to; see Fig. 4.38. This can be based on the service type, time of day, geographical requirements (choosing a physically close partner), etc. This concept brings several advantages and limitations as well.

In this scenario, cloud operators have strict oversight of their Arrowhead clouds and can easily log the interactivity, set up billing, and access control between operators (fits business purposes). It is also highly secure: no need for authentication, identity control, or trust management in any phase. However, there is the need for manual editing of configuration and hence there is absolutely no option for scaling automatically.

The second approach is still based on per-transaction polling of neighbours, but lets the Gatekeepers automatically detect and poll their neighbourhood. The ad hoc peer-to-peer detection of neighbours can provide adap-

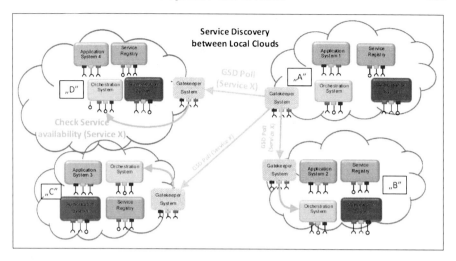

FIGURE 4.38
Service discovery in a hard-wired inter-cloud scenario.

tivity for volatile environments (e.g., moving cars looking for charging stations along the road). However, it will also pose a high administrative overhead (for Gatekeeper components) and cause scaling issues. This methodology will also present a limited list of available resources to the local cloud (just from the current contacts). There are several problems to solve here, such as the tracking, trust management, and authentication of a rapidly changing neighbourhood.

The third approach is pointing towards the creation of a dedicated inter-cloud system, where the "demand and supply" of inter-cloud requests can meet. The main purpose of this is to take off the computational overhead from the Arrowhead cloud Gatekeepers and local core systems in a stockmarket, like environment; see Fig. 4.39. This centralised entity could provide global matchmaking for the participants of the Cloud of Clouds and could also centralize resource allocation and trust management. This can be achieved by tracking the reliability of the parties and providing the requester clouds with only the globally optimal partnering cloud.

4.3.7.2 Inter-cloud negotiations

After the requesting cloud decides for a partnering cloud, the Gatekeepers of these clouds will have to perform negotiations to settle the different aspects of the transaction. At the end of this phase, the requesting system gets a token that will authenticate and navigate it in the partnering cloud, using the local ServiceDiscovery service. During this process, the following issues are to be handled:

1. protocol negotiations (Arrowhead Framework version and protocol matching)

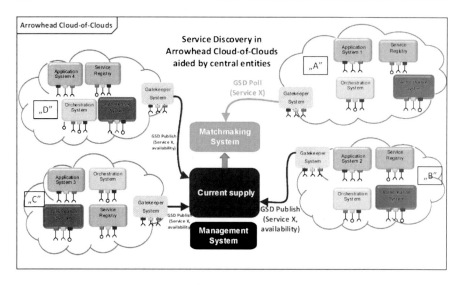

FIGURE 4.39
The Cloud of Clouds concept.

2. mutual authentication and identity checking of the parties

3. actual admission control (authorisation and resource allocation for the transaction in the clouds)

4. establishing a secure connection between the Gatekeepers, therefore the Arrowhead clouds (creating the data path)

5. sharing the local ServiceDiscovery data of the partnering cloud (addressing the partnering cloud from the outside)

6. injecting temporary authorization information into the Authorisation system in the partnering cloud (access control of the "foreigner" system)

7. the partnering cloud Gatekeeper issues a token for the requester with temporary entry and shares it with the requesting Gatekeeper

8. the requesting local cloud forwards this token to the requesting system,

9. the system connects to the producer using the partnering cloud's core Systems

4.3.8 Translation system

The Translation system provides the transparency capability to Arrowhead local clouds. The Translation system hosts on-demand proxy capability as transient service endpoints. That is, the Translation system can be requested to mimic a particular service provider and consumer pair whose service contract does not match. This is shown in Figure 4.40.

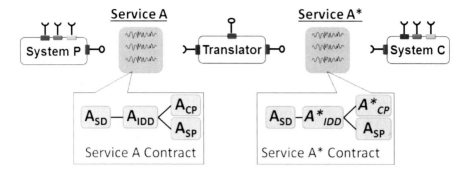

FIGURE 4.40
Arrowhead service contract mismatch.

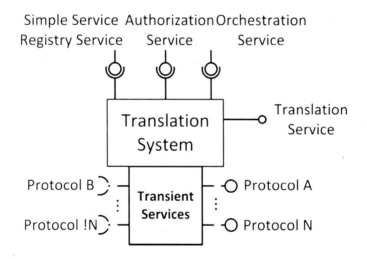

FIGURE 4.41
Translator service interface architecture.

As shown in Figure 4.41, the Translation system consumes the ServiceRegistry service and the Authorisation service and provides a Translation service. The interfaces used for actual translation are transient services and not registered to the service registry. Thus they are dedicated to the consuming systems making the request.

The translation system currently supports translation between HTTP, CoAP, and MQTT. Figure 4.42 shows the block diagram of the translator. Each protocol has been implemented as a spoke segment which connects to any

other spoke segment. Through the use of an intermediary structure new protocols can be introduced through the implementation of a service provider spoke
and a consumer spoke. Then, based on the protocol translation requested, the
hub will connect the appropriate protocol spokes. Thereby achieving efficiency
of direct protocol-to-protocol translation while only requiring implementation
of translation to/from the intermediary structure.

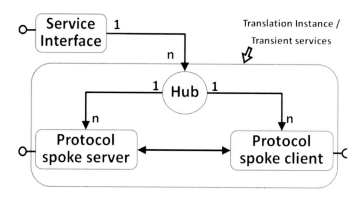

FIGURE 4.42
Translator block diagram.

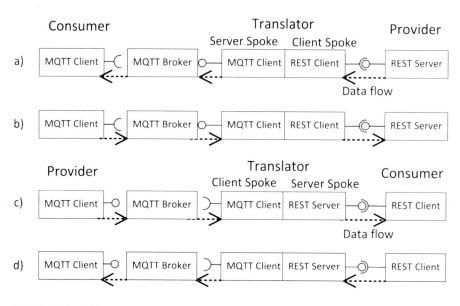

FIGURE 4.43
MQTT translator operational states.

   The translation system architecture additionally allows for evolution of more advanced translation as each spoke can be cascaded as segments. With the addition of semantic, security, or encoding segments, it would be possible to fully support translation between domain-specific communication.

To overcome differences in protocol interaction patterns, a protocol spoke must 1) not rely on any behaviours from a paired spoke and 2) must take proactive behaviours depending on provider/consumer orientation. For example, in Figure 4.43, there are four operational modes for a translator bridging an MQTT spoke and a REST (CoAP/HTTP) spoke. This is bridging RESTful communication with a publisher-subscriber pattern. In Figure 4.43-a and 4.43-b the MQTT client needs to send/receive data to/from the REST server. In Figure 4.43-c and 4.43-d the REST client would like to send/receive data to/from the MQTT client.

The Translation system consumes the service registry and the authorisation service, and produces only a single discoverable service. The translation system dynamically creates service endpoints; however, these are not registered with the service registry as they are only available for use by specific systems which are notified directly by the Orchestration system.

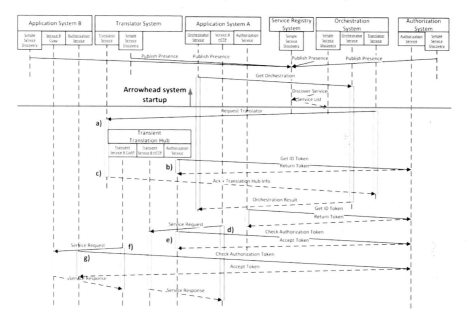

FIGURE 4.44
Translator sequence diagram.

4.3.8.1 Translation services

The Translation service is used to instantiate new translation instances. The interface definition can be seen in Figure 4.45. It takes the service provider service registry record and the service consumer type reference.

```
<function name="get translator">
  <URI name="http://URI:PORT/translator">
    <method name="POST"/>
    <payload>
      <[Mandatory]providerName="string" />
      <[Mandatory]providerAddress="string" />
      <[Mandatory]providerType="string" />
      <[Mandatory]consumerName="string" />
      <[Mandatory]consumerType="string" />
      <[Optional] Event Service name="string" address="string" />
      <[Optional] Quality of Service record>
        < latency="string" />
        < throughput="string" />
        < jitter="string" />
        < packetloss="string" />
      </[Optional] Quality of Service record>
    </payload>
    <response]>
      <Status Code 200>
        <[Mandatory]translatorId="int" />
        <[Mandatory]translatorAddress="string" />
      </Status Code 200>
      <Other Status Codes>
        <no payload in this response />
      </ Other Status Codes >
    </response>
  </URI >
</function>
```

FIGURE 4.45
Translator interface description

Based on this information the translation service will create a new translation hub which will host the protocol spokes. One spoke will make service invocations to the specified service provider and one spoke will mimic the service provider and await service requests from a service consumer. The service endpoint, HTTP/CoAP URL or MQTT broker+topic, will be returned to the service invoker. Each new request to the translator will create a new translation hub.

For the latest update on the Arrowhead Framework and the local automation cloud systems and services, please consult the Arrowhead Framework official wiki [9].

Bibliography

[1] P. Mockapetris, "Domain names - concepts and facilities," IETF, Tech. Rep., 1987.

[2] ——, "Domain names - implementation and specification," IETF, Tech. Rep., 1987.

[3] R. Elz and R. Bush, "Clarifications to the DNS specification," IETF, Tech. Rep., 1997.

[4] A. Hubert and R. van Mook, "Measures for making DNS more resilient against forged answers," IETF, Tech. Rep., 2009.

[5] S. Cheshire and M. Krochmal, "DNS-based Service Discovery," IETF, Tech. Rep., 2013.

[6] "Extensible Markup Language - XML," 2016. [Online]. Available: https://en.wikipedia.org/wiki/XML

[7] D. Peintner and S. Pericas-Geertsen, "Efficient XML interchange (EXI) primer," W3C, Tech. Rep., 2014.

[8] C. Jennings and Z. Shelby, "Media types for sensor markup language (SenML) draft-jennings-senml-10," IETF, Tech. Rep., 2013.

[9] (2016) Arrowhead framework wiki. [Online]. Available: https://forge.soa4d.org/plugins/mediawiki/wiki/arrowhead-f/index.php/Main_Page

[10] "Iso/iec 81346 industrial systems, installations and equipment and industrial products – structuring principles and reference designations," ISO/IEC, Tech. Rep., 2009.

[11] O. Carlsson, D. Vera, J. Delsing, B. Ahmad, and R. Harrison, "Plant descriptions for engineering tool interoperability," in *Proceedings of IEEE INDIN 2016*, 2016.

[12] O. Carlsson, C. Hegedűs, J. Delsing, and P. Varga, "Organizing IoT Systems-of-Systems from standardized engineering data," in *Proceeding IECON 2016*, Florence Italy, Oct. 2016.

[13] M. Albano, L. Ferreira, and J. Sousa, "Extending publish/subscribe mechanisms to SOA applications." in *Proceedings of the 12th IEEE World Conference on Factory Communication Systems (WFCS)*, Aveiro, Portugal, May 2016.

[14] L. L. Ferreira, M. Albano, and J. Delsing, "QoS-as-a-Service in the local cloud," in *Proceedings of SOCNE 2016, in conjunction with ETFA 2016*, Berlin, Germany, Sep. 2016.

[15] M. Albano, R. Garibay-Martínez, and L. L. Ferreira, "Architecture to support Quality of Service in Arrowhead systems," in *Proceeding of IN-FORUM 2015*, Covilhã, Portugal, Sep. 2015.

[16] K. Nichols, S. Blake, F. Baker, and D. Black, "Definition of the Differentiated Services Field (DS Field) in the IPv4 and IPv6 Headers," RFC 2474 (Proposed Standard), Internet Engineering Task Force, Dec. 1998, updated by RFCs 3168, 3260. [Online]. Available: http://www.ietf.org/rfc/rfc2474.txt

5

Application system and services: Design and implementation - A cookbook

Jerker Delsing
Luleå University of Technology

Michele Albano
ISEP, Polytechnic Institute of Porto

Luis Ferreira
ISEP, Polytechnic Institute of Porto

Fredrik Blomstedt
BnearIT

Per Olofsson
BnearIT

Pal Varga
AITIA

Federico Montori
University of Bologna

Fabio Viola
University of Bologna

CONTENTS

5.1 Introduction

In previous chapters local automation clouds and a SOA based architecture supporting the design and implementation of IoT based automation systems were discussed. This chapter is devoted to design and implementation of application services:

- Design of an Arrowhead Framework system

- Implementation of such a system and its services

- Interoperability test

5.2 Application service design

This section will discuss the design of an automation application system and associated services. For this purpose, we will make use of the simple control loop example addressing the level in a flotation tank.

5.2.1 Control application design

The application scenario is a flotation tank in an ore separation plant; cf. Figure 5.1. To operate well, the level in the flotation tank should be above 75%, but also always be below 90% to avoid any spillage.

FIGURE 5.1
An ore flotation tank with overflow of liquid with enriched ore.

For this purpose we have a flotation level sensor measuring the liquid level with a resolution of 0.5 centimeters. The tank itself is 2 meters high. There is an outflow of liquid with aggregated ore. The control task is to keep that outflow at a value of 50 ± 5 l/min. The tank is fed by a pump attached to a speed variator to regulate the flow.

The level sensor transmits level information to a controller which in turn produces set-point data to the pump speed control. The control scenario set up in a local cloud then looks like Figure 5.2. In the context of an Arrowhead Framework local cloud, it is considered that the level sensor, the controller and the speed variator each will host an Arrowhead software system producing or consuming a LevelData and a PumpSpeed service.

According to the Arrowhead Framework documentation guidelines, the information regarding the overall system description, the System of Systems, shall be presented in the SoSD document. The above description is what should be at the heart of the SoSD document.

The individual systems of the control loop system should each be described in a SysD document. For the control loop this breaks down to three SysD documents:

- Level sensor system with its produced and consumed services:

  - Produces the LevelData service

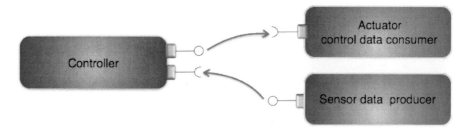

FIGURE 5.2
A control loop consisting of three systems interacting via services to create a
closed control loop.

    – Consumes ServiceDiscovery service

    – Consumes AuthorisationControl service

    – Consumes Orchestration service

- Pump speed controller system with its produced and consumed services:

    – Consumes the LevelData service

    – Produces the PumpSpeed service

    – Consumes ServiceDiscovery service

    – Consumes AuthorisationControl service

    – Consumes Orchestration service

- Pump system with its produced and consumed services:

    – Consumes the PumpSpeed service

    – Consumes ServiceDiscovery service

    – Consumes AuthorisationControl service

    – Consumes Orchestration service

Next is the detailed definition of the produced application services: Level-
Data and PumpSpeed. According to the Arrowhead Framework documentation guidelines, the description of these services shall be provided in Service
Description (SD documents), one document per produced service. The core
content of the SD documents could look like:

- The LevelData service will provide liquid level data in the flotation tank
  with a range from 0 to 2 meters and a resolution of 0.5 centimeters. Level
  data will be provided at least every 1 second.

- The PumpSpeed service will provide a command to increase or decrease
  the pump speed every time the liquid level has changed 2 centimeters (the
  control algorithm itself will not be described).

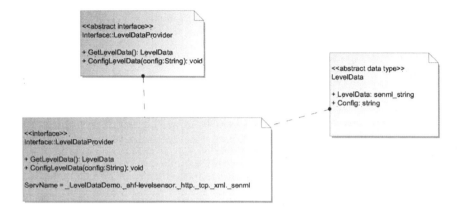

FIGURE 5.3

The LevelData service interface with methods and data types.

Finally, the interface, methods and datactypes of the service will be defined as presented in the Interface Design Description (IDD, document).

Figure 5.3 reflects the interfaces, methods and data types for the Pump-Speed service. This type of graph should be complemented with text definitions of:

- Interfaces and its

- Methods and its

- Data types

The next level is the definition of what technology that is used for the implementation of the above described service. This will be described in the Communication Profile (CP, document).

In this case both the PumpSpeed and the LevelData services are using the CoAP service protocol, the UDP protocol for the transport layer. To protect the payload IPSec [1] was selected as the security mechanism.

Finally the payload data semantics should be defined. This is done in the Semantics Profile, SP document.

Both the PumpSpeed and the LevelData use SenML semantics. As an example, a simple SenML data structure was decided on for the LevelData instance:

```
{ " e ": [
    { " n ": " level ", " u ": " m ", " t ": 0, " v ":1.85 },
  ],
  " bn ": " http:// arrowhead . eu/ level17 /"}
```

The detailed use of SenML semantics or of any other semantics is beyond the scope of this book. Please refer to the SenML documentation [2] or other semantics documentation for further details.

In many cases the CP and SP documents will become the same for many services.

Based on this service definition, it should now be possible for a reasonably experienced programmer to implement the produced Leveldata and Pump-Speed services as parts of the systems residing in the LevelSensor device and the PumpController device.

5.2.2   Control service local cloud administration

The next step in the design of the system interaction and implementation is to have the systems register their services in the local automation cloud to enable the overall application execution. This is what is called administration in the frame of the Arrowhead Framework.

From this local cloud administrative perspective, there are two types of roles a system can take:

- Service producer

- Service consumer

The necessary administrative capability of a system is outlined below. The outline indicates which service interaction code will be necessary for most systems.

For the case of X.509 certificates and the usage of the AA-Authorisation system, it is here assumed that certificates have been distributed by a trusted part, to the involved systems and that the Authorisation system has certificates corresponding to those systems and devices expected to be authenticated within the local cloud.

The specific authentication process is technology dependent. For the case of REST, authentication is made through the hand shaking mechanism used by SSL.

5.2.2.1   Service producer design

In its simplest form a service producing system has to publish its services in the ServiceRegistry system of the local cloud. This is done by consuming the ServiceDiscovery service produced by the ServiceRegistry system. A sequence diagram presenting the process is provided in Figure 5.4. The service producer will use the ServiceRecord interface of the ServiceDiscovery to store its information. ServiceRecord data are encoded in XML or JSON. Their semantic is defined as indicated in Section 3.2.6. The following is an example of a service provided by a temperature sensor:

- Endpoint: e.g., *path/port/*,e.g., *172.28.243.198/ahf/temp17/8077*

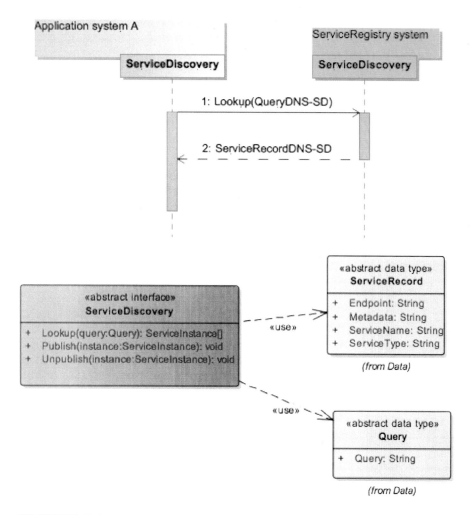

FIGURE 5.4
Sequence diagram for the simplest form of service registration to the ServiceRegistry system. Also shown is the ServiceDiscovery interface and related ServiceRecord on what information about a service the ServiceRegistry can store.

- Metadata: e.g., $\backslash\_xml=1\backslash\_exi=0\backslash\_senml=1md-path=172.28.243.198/ahf/temp17/8077/meta-data$

- ServiceName: e.g., $\_Temp17$'

- ServiceType: e.g., $\_ahf-temp.\_http.\_tcp$'

For details on this please see Section 4.2.1

When a service request is received the producing system has to check with the Authorisation system whether the requesting system is authorised to consume the service. A sequence diagram for such a check is found in Figure 5.5.

This constitues the minimum capability of the service producing system and its interaction with the mandatory core services. This interaction can of course be more complex, if, for example, higher security is demanded by requiring authorisation of all these interactions, though the additional code should be quite small. If other core systems should be utilised, yet additional complexity and code is added. This is omitted here for simplicity.

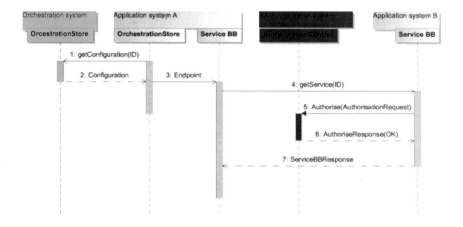

FIGURE 5.5

Sequence diagram for the interaction between an application system A and the Orchestration system and the Authorisation system to enable the consumption of a service BB produced by the application system B.

5.2.2.2   Service consumer design

For a service consumer the interaction is more elaborate. First the consuming system has to be orchestrated by requesting the orchestration information by consuming the OrchestrationStore service. The interface used is getConfiguration, upon which the consuming system will receive the endpoints of the services to consume. For each endpoint there is associated data defining the SLA of the produced service. Here protocols, interfaces, metods, datatypes, data encoding, semantics and eventual compression is specified together with QoS and orchestration priority data.

The design time approach is to use the IDD, CP, and SP documentation to design the consuming system code for a specific service. These documents hold all information on protocols, interfaces, metods, data types, data encoding, semantics, and eventual compression used by the produced service.

More attractive is to enable the run-time creation of a consumer that can use the SLA of the producer. This can be achieved by fetching the SLA of the service to consume from the ServiceRegistry. The following SLA data can be found is the ServiceRegistry:

- Protocol, e.g., CoAP is found in the EndPoint data `coap://192.168.10.123/service/.....`

- Interface, methods, data types are found from metadata, e.g., *wadl=link*

- Semantics is found in metadata, e.g., *sem=senml*

- Encoding is found in metadata, e.g., *encode=xml*

- Compression is found in metadata, e.g., *comp=exi*

Based on this it should be possible to implement code that enables a consumer system to consume the provided service through its interfaces using the available methods and data types.

If protocol, encoding, and semantics translation is required this should be detectable by the Orchestration system based on data in the ServiceRegistry and the SystemRegistry. If translation is needed the Orchestration system will invoke the Translation system to perform the necessary translations.

In this way a very wide interoperability can be provided through the translation between protocols, encodings, compressions, and semantics. Further autonomous handling of arbitrary service interfaces, methods, and data types to be consumed is achievable using approaches like, e.g., WADL or HATEOAS [3, 4] .

Once the consumer system has orchestration data, it can call the service endpoint to access the service. While receiving the service request, the service producer will check if authentication of the consumer is required. If it is required and if the appropriate security key is provided by the consumer, then the service will provide its response. Otherwise the consumer will be notified that access to the service has been refused. The consumer may therefore have to request an authentication key from a trusted distributer of authentication keys to successfully consume the service.

5.3 Demo system

To ease the discovery and exploration of the many capabilities of the Arrowhead Framework, the Arrowhead Framework Wiki ([5]) provides access to a Demo system designed for this purpose: `http://forge.soa4d.org/arrowhead-f/4_Howtoimplementapplicationsystems/4_Examplesys-tems/Demosystem/`

The Arrowhead Framework wiki also provides the related documentation (SysD, SD and IDD documents).

The Demo system is a reference implementation in Java of an Arrowhead compliant system. The Demo system demonstrates use of the base Arrowhead Framework core services:

- ServiceDiscovery

- AuthorisationControl

- OrchestrationStore

The use of these services is illustrated around a basic provider and consumer system, with Temperature and Range services.

The implementation offers these services via the REST_WS-TLS-XML Communication Profile. An overview is shown in Figure 5.6.

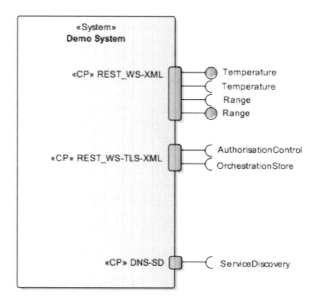

FIGURE 5.6
Demo system service interfaces.

The produced services are

- Temperature

- Range

The consumed services are

- ServiceDiscovery

- AuthorisationControl

- OrchestrationStore

- Range

- Temperature

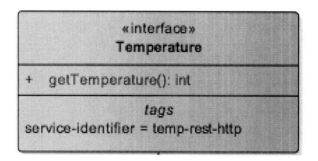

FIGURE 5.7
Temperature service interface.

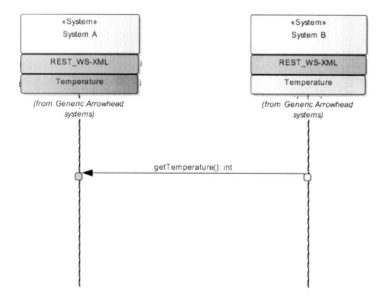

FIGURE 5.8
GetTemperature message sequence diagram.

Let's concentrate on the Temperature service. The service interface is shown in Figure 5.7. Thus requesting the service is made with the method:

- GetTemperature()

Following the message sequence in Figure 5.8.

It's strongly suggested to study the code of this demo example to get a further understanding of how to implement your own Arrowhead compliant services.

5.4 Implementation — Simple examples

The example provided here is a simple demonstration program to set up a service provider and the resources used, based on REST and the http protocol.

The source code given here is also available in the "How to implement" section of the Arrowhead Framework Wiki.

The simple example consists of a REST-WS provider and a consumer that can interact with each other. They use the ServiceRegistry for registration and lookup of the services. It also provides illustration of the way to consume the AuthorisationControl service. Besides showing the interaction with the AuthorisationControl service, the example also shows how to make a request using TLS and demonstrates the usage of a Java keystore.

5.4.1 Simple producer —- Code example

This section describes the code of a system producing a simple service and registers it within the service registry.

```
// First create your provider class based on the
// basic REST ( http) provider
public class SimpleProvider extends BaseProviderREST\_WS

// Initiate Service Discovery for use to publish your
// resource
sd = new ServiceDiscoveryDnsSD ();

// Initiate your service , create your REST resource
// endpoint and start the underlying inherited
// base functionality
SimpleProvider simple = new SimpleProvider ( log , sd );
Resource = new SimpleProviderResourceREST\_WS(" Simple data
here ");
endpoint = new HttpEndpoint ( serviceDiscovery . getHostName (),
25252 , "/ data /");
simple . start ();

// Your service is now up and running!
// Publish your provider in the service registry and make
// your service visible and possible to discover in the
// service registry by other systems.
simple . publish ();

// When the system is terminated , remove your system
// from the service registry and stop your service ' s
// resources
simple . unpublish ();
simple . stop ();
```

5.4.2   Simple consumer — Code example

Here is provided code for a system that will consume the simple service produced above.

```
// Create your service class
public class SimpleConsumer
// Declare and initiate your Service Discovery service
// API to access the Service Registry
ServiceDiscovery sd = new ServiceDiscoveryDnsSD();

// List all services in the Service Registry
List< ServiceIdentity> services =
sd. getServicesByType( SimpleProvider. DNS\ _SD\ _TYPE);

// Create a Consumer REST ( factory) to use for connecting
// to the service provider to request data
ClientFactoryREST\ _WS clientFactory =
new ClientFactoryREST\ _WS();

// Search through all services for the required Service
// Type, in our case the HTTP ( REST) service.
// Note: In a fully functional system
// the Orchestration System should do this for!
// Then we should as a consuming system connect to the
// Orchestration System and fetch the information for
// which service to connect to!

for ( ServiceIdentity si : services) \{
ServiceInformation serviceInfo =
sd. getServiceInformation( si, HttpEndpoint. ENDPOINT\ _TYPE);
if ( serviceInfo != null)\{
HttpEndpoint httpEndpoint =
( HttpEndpoint) serviceInfo. getEndpoint();\}

// Create the Consumer target client instance to the
// ServiceProvider resource endpoint
Client client = clientFactory. createClient( false);
WebTarget target;
try {
target = client. target( httpEndpoint. toURL(). toURI());
URL targetUrl = target. getUri(). toURL();\}

// Make the call to the Resource
SimpleData response =
target. request( MediaType. APPLICATION\ _XML\ _TYPE).
get( SimpleData. class);

// Close the consumer connection to the provider resource
client. close();
```

5.4.3   Authorisation service consumer

Here is demonstrated the use of the Authorisation service and a consumer using HTTPS. As the service uses HTTPS, this client requires a valid client certificate in the key store, and the server certificate or a CA certificate of the Authorisation service in the trust store. Additionally, as the system is validating hostnames, the name resolution must work on the underlying system.

```
public   class   AuthorisationServiceConsumer  \{
//  Setting  up  the  DN  to  be  queried  authorisation  for.
//  This  is  normally  taken  from  a  client  certificate.
String  DN = " C= SE, ST= Sweden, L= Lulea, O= BnearIT,
OU= Development, CN= SomeToBeAuthorised ";
ClientFactoryREST\ _WS  clientFactory =
new  ClientFactoryREST\ _WS( trustfile, trustpass,
keystorefile, keypass);\}
ServiceDiscovery  sd = new  ServiceDiscoveryDnsSD ();

List< ServiceIdentity> authServices = sd. getServicesByType
( AuthorisationServiceTypes.
REST\ _WS\ _AUTHORISATION\ _CTRL\ _SECURE);

//  Loop  through  all  service  instances  and  make  a  call
//  on  each  authorisation  provider.
for ( ServiceIdentity  si : authServices) \{
  ServiceInformation  serviceInfo =
. getServiceInformation( si, HttpEndpoint. ENDPOINT\ _TYPE);

  AuthorisationControl  authControl =
new  AuthorisationControlConsumerREST\ _WS
(( HttpEndpoint) serviceInfo. getEndpoint (), clientFactory );

String  serviceType = AuthorisationServiceTypes.
REST\ _WS\ _AUTHORISATION\ _CTRL\ _SECURE;

String  serviceInstance = si. getId ();

boolean  isAuthorised = authControl. isAuthorized(
                DN,
                serviceType,
                serviceInstance );

System. out. println(" Performed  authorisation  check  on  "
+ serviceInfo + "  for  '" + DN +"'. The  result  is  " +
isAuthorised );\}
```

5.5   Deployment of a local cloud

This section provides step by step instructions on how to install a basic Arrowhead Framework local cloud. It is assumed that you are running a Linux installation. The example is based on an freshly installed CUSTOM-CENTOS [6] with the environment of

- **hostname**: bedework.arces.unibo.it

- **domain**: arces.unibo.it

- **IP address**: 137.204.143.20

- **gateway**: 137.204.143.254

Next DNS, NTP, and Glassfish need to be installed and configured as per below. This is then followed by the installation of the Arrowhead Framework mandatory core systems and some useful tools for the administration of an Arrowhead Framework local cloud. Everything needed for an installation is found in the CUSTOM-CENTOS.iso located at the Arrowhead Framework wiki.

5.5.1   DNS configuration

The example is using yum (an rpm installer [7]). First install the DNS server Bind:

```
# yum install bind -y
# chkconfig named on
# cd /root/installation/zone-file-gen
# ./interactive_setup.sh
```

The last command requires some input from the user: you can confirm the default configuration for each parameter.

Now the configuration for the DNS server must be set up. So, given our network configuration, the files to be edited are

- arces.unibo.it

- 143.204.137.in-addr.arpa

So the first file should look like this one:

```
$ORIGIN  .
$TTL  1 D
arces. unibo. it  IN  SOA  ns. arces. unibo. it.
root. arces. unibo. it.  (
                                        1       ;  serial
                                        1 M     ;  refresh
                                        1 H     ;  retry
                                        1 W     ;  expire
                                        3 H  )  ;  minimum
                    NS  ns. arces. unibo. it.
$ORIGIN      arces. unibo. it.
ns           A    137.204.143.20
bedework     A    137.204.143.20
server       A    137.204.143.20


$ORIGIN    _dns- sd. _udp. arces. unibo. it.
b            IN  PTR  srv. arces. unibo. it.
db           IN  PTR  srv. arces. unibo. it.
lb           IN  PTR  srv. arces. unibo. it.
r            IN  PTR  srv. arces. unibo. it.
dr           IN  PTR  srv. arces. unibo. it.
```

while the second file should look like the following:

```
$TTL  1 D

@ 1 D  IN  SOA  ns. arces. unibo. it.
root. arces. unibo. it  (1  4 h  1 h  1 w  1 h)

@ 1 D  IN  NS  ns. arces. unibo. it.

20 1 D  IN  PTR  ns. arces. unibo. it.

20 1 D  IN  PTR  bedework. arces. unibo. it.
```

Now edit the configuration of /etc/resolv.conf to maintain only the following line:

```
nameserver 137.204.143.20
```

Let's make immutable the file /etc/resolv.conf:

```
# chattr +i /etc/resolv.conf
```

In this way the DNS server used to resolve hostname will be the DNS server installed on this machine.

> **IMPORTANT:** This is a critical step! A different configuration of the file /etc/resolv.conf will produce exceptions during the installation of the core services!

5.5.2 NTP

To enable time synchronisation, NTP [8] should be loaded.

```
# cd /root/installation
# ./install-ntp-server.sh
```

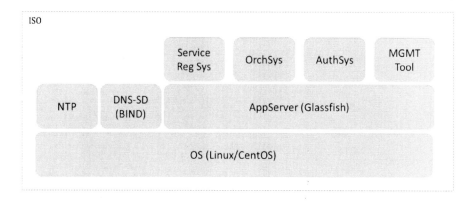

FIGURE 5.9
Glassfish supports the implementation of all three mandatory core systems of the Arrowhead Framework.

5.5.3 Glassfish

Now it is time to install a Glassfish server [9]. Glassfish is used by the implementations of all three mandatory core systems, see Figure 5.9 . This one of the most delicate steps of the installation process. Before the installation be sure that the configuration of /etc/resolv.conf is correct.

```
# cd /root/installation
# ./install-glassfish.sh
```

5.5.4 Core systems

Next the Arrowhead Framework mandatory core systems are installed.

```
# yum install ServiceRegistry -y
# yum install Authorisation -y
# yum install Orchestration -y
```

5.5.5 Password-less login

Next let's enable automatic login.

```
# ./install-ssh-keys.sh bedework.arces.unibo.it
```

5.5.6 Management tool

The management tool will be installed by the commands

```
# yum install ManagementTool -y
# yum install TestTool -y
```

5.5.7 Test the configuration

Point your browser on:
`http://bedework.arces.unibo.it:8080/managementtool`

Login and check if you can retrieve the list of the active core services. Also check if the logging functionalities and the certificates tabs work properly.

If succesful you should now have a device hosting the mandatory core systems of the Arrowhead Framework and two of the most valuable tools, the Management tool and the Test tool, to get started with Arrowhead Framework.

5.6   Arrowhead Framework tools

5.6.1   Management tool

The Arrowhead Framework provides a simple system management tool to edit and assign orchestration rules. The tool provides a graphical user interface that allows a user/operator to create connection rules (orchestration) for systems (i.e., for System A and System B).

The major purpose of the Management tool is to help an operator to create a composition of System of Systems that fulfils basic needs. It is aimed providing an easy and flexible way to reconfigure and distribute new composition configurations to operators and responsible personnel.

FIGURE 5.10
The user interface of the Arrowhead Framework Management tool.

The tool uses the service OrchestrationManagement to communicate with the Orchestration system and its stored connection rules. The ServiceDiscovery service is used to list possible services to set connection rules for a system.

The tool can build a picture of the current state of the actual system collaboration and dynamics in run time. This is done with the support of the OrchestrationCapability service. Here is provided information on which service producers are consumed by which service consumers within the local cloud. The main interface of the Management tool is depicted in Figure 5.10. More documentation and code for the Management tool is available at the Arrowhead Framework wiki.

5.6.2   Test tool

The Arrowhead Framework Test tool allows developers to verify interoperability between Arrowhead Framework compliant systems. The Test tool enables the verification of:

- Produced service and its interfaces, methods and data types.

- Service consumption of a specific service.

- Orchestration of consumption of services to be tested.

Arrowhead Framework Test tool is a stand-alone application used as a test utility to test, validate and simulate service interfaces. It is possible for each developer to download the tool and deploy it for local tests. Code and documentation is from the Arrowhead Framework Wiki. Figure 5.11 shows the user interface of the test tool.

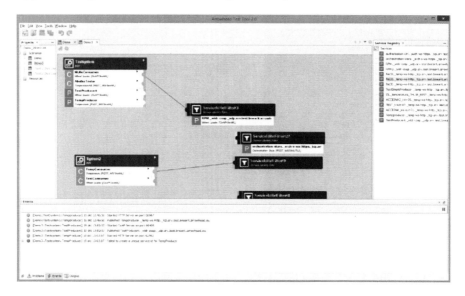

FIGURE 5.11

The user interface of the Arrowhead Framework Test tool.

Currently only a few application service can be tested using the test tool currently available from the Arrowhead Framework wiki. Test of additional produced application services requires an implementation of consumer code that can address the provided interfaces, methods, and associated data types. The test tool will implement the service type according to appointed and approved IDD. A scripting support is provided within the tool. In the future it is expected that the test tool will be able to autonomously detect the capabilities of the produced services and provide test capability thereof.

Bibliography

[1] "Internet protocol security (ipsec)," 2016. [Online]. Available: https://en.wikipedia.org/wiki/IPsec

[2] C. Jennings, Z. Shelby, J. Arkko, and A. Keranen, "Media types for sensor markup language (senml) draft-ietf-core-senml-00," IETF, Tech. Rep., 2016.

[3] M. Hadley, "Web application description language," W3C, Tech. Rep., 2009.

[4] Wikipedia, "Hateoas — wikipedia, the free encyclopedia," 2016, [Online; accessed 2-June-2016]. [Online]. Available: https://en.wikipedia.org/w/index.php?title=HATEOAS&oldid=715478952

[5] (2016) Arrowhead framework wiki. [Online]. Available: https://forge.soa4d.org/plugins/mediawiki/wiki/arrowhead-f/index.php/Main_Page

[6] Wikipedia, "Centos — wikipedia, the free encyclopedia," 2016, [Online; accessed 6-June-2016]. [Online]. Available: https://en.wikipedia.org/w/index.php?title=CentOS&oldid=723592579

[7] ——, "Yellowdog updater, modified — wikipedia, the free encyclopedia," 2016, [Online; accessed 6-June-2016]. [Online]. Available: https://en.wikipedia.org/w/index.php?title=Yellowdog_Updater,_Modified&oldid=719079512

[8] ——, "Network time protocol — wikipedia, the free encyclopedia," 2016, [Online; accessed 6-June-2016]. [Online]. Available: https://en.wikipedia.org/w/index.php?title=Network_Time_Protocol&oldid=723622780

[9] ——, "Glassfish — wikipedia, the free encyclopedia," 2016, [Online; accessed 6-June-2016]. [Online]. Available: https://en.wikipedia.org/w/index.php?title=GlassFish&oldid=720888274

6

Engineering of IoT automation systems

Oscar Carlsson

Midroc Automation AB

Daniel Vera

Fully Distributed Systems Ltd.

Eduardo Arceredillo

Fundacion Tekniker

Markus G. Tauber

Fachhochschule Burgenland GmbH

Bilal Ahmad

University of Warwick

Christoph Schmittner

Austrian Institute of Technology

Sandor Plosz

Austrian Institute of Technology

Thomas Ruprechter

Infineon Technologies AG

Andreas Aldrian

AVL List GmbH

Jerker Delsing

Luleå University of Technology

CONTENTS

6.1   Introduction

The acceptance of new technology depends to a high degree on usability for the end users. As the continued usability of technical systems depends on the engineering support throughout the life cycle, the support for engineering and continued operation is vital for successful acceptance of new technology. As the Arrowhead systems are envisioned to cater to a wide array of users in different domains of our society, there will be a need for different methodologies depending on both the requirements in different areas as well as the skills expected from the different users.

Across the domains covered by Arrowhead partners a select number of recommended methodologies have been identified and described. To further illustrate the situations where the different methodologies may be suitable, a number of scenarios have been provided. The scenarios range from deployment of numerous sensors at a minimal engineering and operation cost, through high-security, industrial applications to replacement of devices due to failure or maintenance operations.

6.2 Engineering of an Arrowhead compatible multi-domain facility

For very large projects involving automation, especially in projects geared towards the process industries, it is common that engineers from different disciplines are required to coordinate the designs of buildings, electrical systems, utilities, piping and instrumentation, process equipment, and automation systems. Most parts used in each individual discipline are either strictly standardised or available as commercially off the shelf (COTS), making the design more of a coordination and composition activity than free design work. The engineering process and its associated information management are described in greater detail by Yang [1].

Throughout the design and engineering phases, different tools will be used for the different disciplines. Most of these tools can be grouped together as computer-aided design (CAD) software. Many of the tools are based on or can export to databases but they are typically not made to be integrated directly. As the existing tools are very capable in their respective fields, it is proposed that they are kept separate and that engineers can keep working using the tools that are best suited for their discipline.

The interoperability of engineering tools is possible today - as long as all partners follow the same standards. However, as the Arrowhead project targets many domains and engineering tools have to cover different aspects and life cycle phases of those domains, there are many standards that could be used depending on the domains and life cycle phase. Even within a single domain there can be many standards available, as illustrated by the comparison of standards by Braaksma et al. [2].

In order to synchronise, it's suggested that the design engineers uses the Arrowhead Framework, throughout the whole process and uses the PlantDescription (Section 4.3.1) as a reference tool for objects and systems that have been identified and how they are related to each other. At the early stages the PlantDescription can help to coordinate their work, as the design initially is often focused on the overarching functionality and names of objects, and responsibilities of subsystems may not be fully decided yet. At this point the PlantDescription can illustrate how many subsystems may be present in each larger section and how these subsystems are identified within each discipline.

If the tools are enhanced to integrate with a PlantDescription system directly, it will allow engineers from different disciplines to populate the design with specific objects and the data can be synchronised between design tools using the PlantDescription. Throughout the design phase, all of the engineering data is still stored and maintained in the formats preferred by the engineering tools in their respective databases. The data in the PlantDescription should be limited to what objects are present in the separate databases, how they are identified, and what their relations are.

6.2.1   Further development of engineering tool interoperability

As many domains already follow standards relating to one or more of the aspects that this task aims to address, it seems unreasonable to make all of them follow one standard. Within the project a number of standards have been identified that all relate to engineering data to some extent. The following list includes a selection of them:

- CAEX (IEC 62424) [3]

- AutomationML [4] (IEC 62714)

- MIMOSA [5]

- OPC UA [6] (IEC 62541)

- IEC 61449 Function blocks [7]

- IEC 81346 Industrial systems structuring principles [8]

- ISO 15926 Industrial automation systems and integration [9]

- fastAPI Swedish building automation [10]

- IEC 61850 Power utility automation [11]

To alleviate the situation with several standards at the same site, there are already some initiatives on specific synchronisation between pairs of complementing standards such as collaboration between AutomationML (IEC 62714) and OPC UA (IEC 62541) [12] as one example and collaboration between ISO 15926 and Mimosa [13] as another.

The Reference Architecture Model Industrie 4.0 (RAMI4.0) status report [14], a product of the German initiative Industrie 4.0 [15], is centred around the standards IEC 62890 for structuring the life cycle and value stream, combined with the two standards IEC 62264 (ISA-95) and IEC 61512 (ISA-88) for structuring the hierarchical levels. At a more detailed level the report suggests a number of standards for different aspects. For implementation of the communication layer, the report suggests OPC-UA; for the information layer IEC 61360 (ISO 13584-42), eCl@ss, Electronic Device Description (EDD), and Field Device Tool (FDT) are suggested. Field Device Integration (FDI) is suggested as integration technology and for end-to-end engineering the report suggests ProStep iViP, eCl@ss, and AutomationML (which uses a topology based on IEC 62424).

Many of the standards concern the parametrisation of the object data, which is important for compatibility between tools within one domain, but as the target here is interoperability between tools from different domains it may be sufficient to focus on a few key parameters for each object and the different relations the objects have among themselves. Once the interoperability reaches the point of complete integration of all engineering data into a single data store, the standard ISO 18876, which establishes an architecture and methodology for integrating industrial data, could be used.

6.3 Component-based engineering methodology

The concepts and paradigms described here form a basis on which tools and methods for supporting component-based manufacturing system life cycle can be implemented. The procedure focuses on the engineering lifecycle of PLC based automation systems and was developed with the automation systems used for automotive engine assembly operations in mind.

In the domain of manufacturing engineering, the concept of component and component-based engineering methodologies aims at providing some level of support for re-use of pre-validated engineering data across engineering programs. The concept of component may be materialised physically by a mechatronic or IoT device integrating mechanical, electronic and software elements. However, a component is more commonly perceived as a data container that encapsulates various aspects of engineering and their corresponding datasets (e.g., mechanical engineering, control data, process data) for a sub part of a system and at a given level of granularity. The engineering methodology then consists in supporting re-configuration and re-use of components in order to accelerate the design of a complete system.

6.3.1 Life cycle dimensions

The life cycle of component-based automation system is defined as the composition of two life cycles that can be concurrent and/or sequential in time:

- Component engineering life cycle

- Component-based system engineering life cycle

The component and system engineering life cycles can be concurrent and/or sequential in time (depending on system and process-related constraints), but should be de-coupled as much as possible by dissociating the engineering life cycle of individual components and the composition of components into a complete system.

The component life cycle will focus on designing a set of individual components (Component Library) whose chosen levels and types of functionalities allow for systems that provide the expected characteristics to be composed. The component engineering life cycle focuses on designing, pre-testing, validating and maintaining individual Components, while the system life cycle focuses on configuration and composing components into a complete system.

The definition of the initial set of components (Component Library Design) and individual components design will evolve through time, based on the knowledge collected through multiple system engineering life cycles. An example in the domain of manufacturing system is the design of a product clamping unit (or clamping component) that might be updated based on the failures records of similar components deployed in various production lines, and/or the mechanical design may be updated to accommodate a wider range of product sizes, and/or a clamping unit and part presence sensor unit may be merged into one single component in order to facilitate the later system configuration task.

6.3.2   Design dimension

The intrinsic nature of systems used in the manufacturing industry (as in most technological domains) requires expertise in several domains of engineering. Typically, automation systems are electro-mechanical systems in which engineering relies on sub-design activities such as electrical, process control, hydraulic, electronic engineering, etc.

For component-based system design, the implication of having several design dimensions has to be considered from both component and system design perspectives. Figure 6.1 shows schematically how design dimensions can result in components of different natures. The nature of a component might be of a single type, i.e., support only one type of functions (e.g., type/function A), or a component might result from the integration of several design dimensions (e.g., electronic and mechanical and software etc).

The choice of building a library based on single-dimension or multidimension component designs, or a mix of both, might be dependent on many parameters and/or system requirements: nature of the system being designed, requirement in terms of component re-usability and system re-configurability, design commonality strategy, technological limitations, etc. However, it is critical for tools and methods to provide some form of support across all design engineering dimensions, whether at the component or system design life cycle

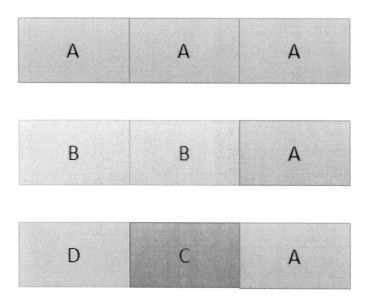

FIGURE 6.1

A component might support only one type of function or might result from the integration of several design dimensions.

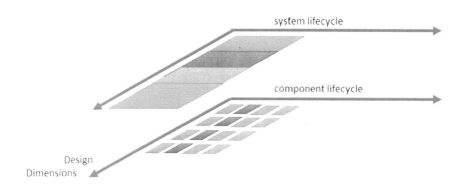

FIGURE 6.2

Design dimensions across system and component life cycles.

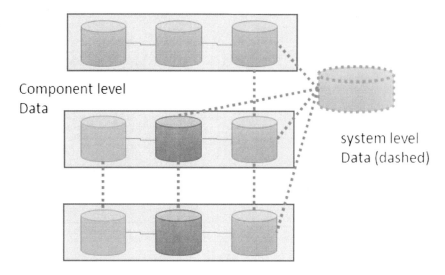

FIGURE 6.3
Component and system data for the design dimensions used, cf. Figure 6.1.

or both. Figure 6.2 illustrates the possibilities with design dimensions across life cycles.

6.3.3   Data model

Modern system engineering activity results in and/or initiates from a set of digital data. Specific design dimensions of a system will be defined by a specific dataset expressed in one or more data formats and generated by specific engineering domains or organisations. The same design dimension might be described using different datasets and data formats (e.g., various CAD formats, different control code, etc.). The dataset and data formats used to describe a system or sub-system are defined as the data model.

For component-based system design, the data model exists at two levels of abstraction: the component and the system levels. At the component level, the data model will reflect the grouping of various design dimensions and define a dataset that includes engineering-specific data expressed in engineering-specific formats, cf. Figure 6.1 and 6.3. The component data model should also include data describing the correlation between various design dimensions (e.g., mechanical actuator position corresponding to a given logic state).

The system level data model will define the data required to achieve functional composition of component data (e.g., process sequence interlock) and any additional data required to support system level design dimension, their definition and functionalities.

Digital data enable computer models to be implemented for virtual design, testing, and validation of both component and component-based systems.

FIGURE 6.4
Gap between engineering and operation phases of the same system.

Digital models and the virtual engineering activity that such models enable are of significant importance when designing engineering tools and methods. Typical virtual engineering tools used in manufacturing system engineering (e.g., PLM) often include a database and data management system, system engineering (i.e., engineering data editing) as well as a set of simulation environments for virtual validation of system design.

Virtual engineering is a critical enabler for achieving efficient design and validation of component systems. Data models and the virtual models used to simulate and validate components and systems at various stages of their life cycle create a new design dimension/life cycle space. One major objective when designing engineering methods and tools is to avoid the discrepancy between the virtual system definition and its physical implementation which typically occurs after system commissioning (physical implementation and deployment), cf. Figure 6.4.

6.3.4 Design guidelines for component-based engineering tools

Following the line of design points made in the previous sections, the following general guidelines for engineering tools and methods can be stated:

- Life cycle aspects specific to component-based system design

  - It is essential to reinforce the difference between component lifecycle (component design) and system life cycle (system configuration)
  - Component and system design should not be integrated, so that system design can be achieved with minimum component re-design
  - Component and system life cycle should be coupled so that system design knowledge can be used to refine the design of component design and/or component library composition
  - Components should be readily configurable pre-tested, pre-validated sub-systems units

- Design dimensions

  - Engineering tools should provide flexibility with regard to the nature, types, and level of granularity of the components that can be designed and subsequently stored in the library

- Engineering tools should provide a clear differentiation between component design and system design workflow and UI

- Engineering tools should reinforce best practices (for component design and system configuration) for the type of system being considered

- Engineering tools should provide functions adapted to specific a design dimension (i.e., engineering domains)

- Data model properties

  - Engineering tools should enable editing and storage of component and system design data in appropriate preferably open formats to enable integration with other engineering tool environments

  - Engineering tools should provide collaborative capabilities in order to enable component and system data editing by different, possibly distributed, organisations

  - Engineering tools should provide seamless integration of virtual and real component and system data editing processes

  - Engineering tools and associated databas management systems should provide the ability to maintain consistency between digital system data and the real system (after commissioning/deployment phases)

  - Engineering tools should reinforce a component data model versus a system data model (access, editing, visualisation, etc).

6.4 Safety and security engineering of IoT automation systems

Modern IoT automation systems are being increasingly connected to the Internet, especially when cloud services are involved. This means the overall system is exposed to cyber security risks, cf. Figure 6.5. As automation systems are used to control industrial equipment, such cyber security threats could have an effect on the operational safety of the overall automation system. Traditionally, the system and especially the safety- and mission-critical part linked to the control of the mechanical industrial device was isolated from the Internet. Hence, while safety and reliability engineering had a long tradition in this domain, cyber security issues have played a minor role.

Therefore, when engineering an IoT automation system, both safety and security and their interplay must be considered and assured in order to guarantee a smooth operation and trust in the system.

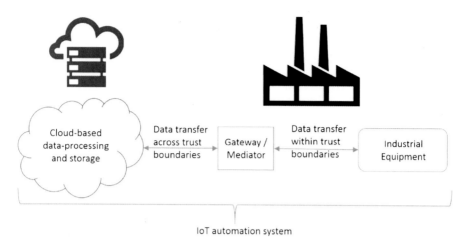

FIGURE 6.5
Basic architecture of industrial IoT automation systems.

6.4.1 General approach

Based on the methods that have been consolidated in the research phase of the project, a safety and security analysis has been made in sync with the engineering process. The methods for both safety and security analysis include

a) identification of the assets of interest for the stakeholders operating the IoT automation system followed by system modelling

b) identifying vulnerabilities and threats (security) and their counterparts in the safety domain (i.e., failure modes)

c) ranking them using risk assessment approaches

A special focus was on threats or failures which threatened cross-domain properties, e.g., security threats which endangered the safety or reliability of operation. The result of risk assessment provides a feedback for system engineers about the security, safety, and reliability of the developed system and defines requirements for next iterations of the security and safety concept.

6.4.2 Performing security analysis

Parts of the methods described for security analysis on IT architectures in an M2M (machine to machine) context have been published [16]. Based on safety and security analysis methods, developed by Austria Institue of Technology, an architectural safety and security analysis on the Arrowhead pilot "smart-engine pilot" was conducted. This section details the security assessment methodology.

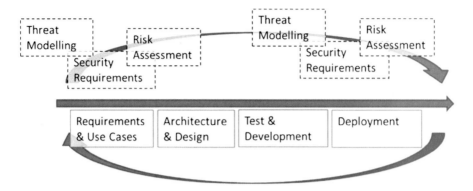

FIGURE 6.6
Security-aware system design life cycle.

FIGURE 6.7
Terms and relations.

The individual architectural security analysis activities need to be conducted iteratively and in parallel to the development process of the entire system. Figure 6.6 captures some of the most important activities considered as part of our security analysis and how this aligns with other activities. Further security analysis related activities, e.g., the design and evaluation/testing of security features, may be added later. Some methods are defined per each of the activity groups above, but methods and standards also exist that capture multiple groups of activities. Another distinction between methods is that some are considering technical aspects in depth. Other methods address high-level technical and organisational issues.

The terms which we will be using in this document and their relations are depicted in Figure 6.7 and explained as follows. Threat is a possibility that a system can receive an attack. An attack which is created by a threat exploits the vulnerabilities of the system. Vulnerabilities make a system more prone to an attack by a threat or make an attack more likely to have some success or impact. The likelihood is the frequency or probability of something to occur. The impact is the result of an attack on the system which may be damage, loss of information or service, manipulation of data, etc.

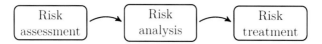

FIGURE 6.8
ISO 27005 Security Risk Management Process.

6.4.2.1 Methods

The security analysis is usually performed as depicted in Figure 6.6, starting with threat modelling, followed by definition of requirements and ending with the risk assessment procedure. This method is along the risk assessment guideline defined in ISO 27005 standard, on which we rely when selecting the security analysis methods. The basic steps to perform risk assessment using the guideline are depicted in Figure 6.8 based on the ISO 27005 standard [17].

According to the ISO guideline, risk assessment consists of risk analysis and risk evaluation. Risk analysis consists of risk identification and risk estimation. The ISO standard was taken under consideration when developing the security analysis methodology used in the pilots.

Risk modelling serves the purpose of "abuse cases identification." "Security Requirements" help to initiate security analyses but can also be a result of it. "Risk assessment" allows linking the probability of likelihood and the effect of a threat becoming an attack to, e.g., determining which counter measure development should be prioritised. During research multiple risk modelling, analysis, and assessment methods have been investigated. The methods which have been consolidated to the security analysis methodology are detailed in the following sections.

Microsoft threat modelling process

The Microsoft threat modelling process originated from secure code writing efforts at Microsoft in the early 2000s. In 2006 the Microsoft security development life cycle was developed [18]. The threat modelling process consists of the steps shown in Figure 6.9.

In the initial step the security objectives need to be defined. Decisions made in this step have a great influence on the following steps in the analysis process. The goal of the next two steps, application overview and decompose application, is to gain a good understanding of the underlying system architecture. As an output of these two steps, architecture diagrams on different levels of detail are created. After having a clear understanding of the system architecture, potential threats to system security and already known vulnerabilities are collected.

The decomposition of the application is normally achieved by developing a data flow diagram (DFD), which consists of processes, data stores, boundaries, and the data flows between them. For identifying threats and vulnerabilities the so-called STRIDE [19] method is applied on each DFD component; -

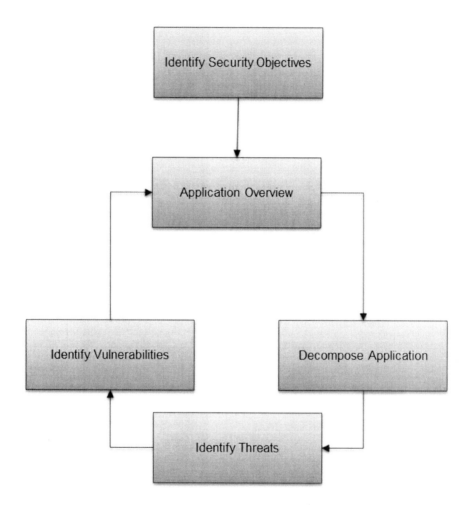

FIGURE 6.9
Microsoft threat modelling process steps.

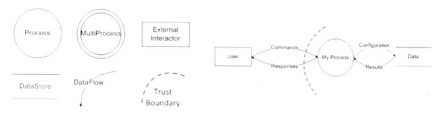

FIGURE 6.10
Modelling objects used in a data flow diagram.

each letter stands for a category of threats: Spoofing, Tampering Repudiation Information disclosure, D.o.S, and Elevation of privilege.

Results

In order to use Microsoft Threat Modelling with STRIDE, one needs to know which assets are concerned. This can be discovered by interviewing system experts. Threats against assets are identified when using STRIDE after decomposing a system in data source, processes, data flow, and interactors in a DFD. The most important DFD components are shown in Figure 6.10 and in a simplistic use case example.

The output or result is a comprehensive list of categorised threats. The identified threats and vulnerabilities can then be ranked with any risk assessment approach to derive the most important threat to deal with.

Usage in Arrowhead

The MS threat modelling process is currently applied to a work package pilot. The methodology presented allows application to other pilots as well. Four cases of high-level architectures have been developed as a starting point for evaluating the security analysis. This means that by categorising the use cases in such a way, vulnerabilities/threat examples can be used between pilots. The four cases are

1. A measurement device talking directly to a back-end system which stores data in a "cloud" or central back-end (called "cloud" for simplicity hereafter)

2. A measurement device talking via a proxy (for multiple devices) at the client side to a back-end system which stores data in a "cloud"

3. A measurement device talking directly to a back-end system which stores data in an external "cloud"

4. A measurement device talking via a proxy (for multiple devices) at the client side to a back-end system which stores data in an external "cloud"

TABLE 6.2
Scale, detectability and recoverability assessment

		Detectability and Recoverability		
		Easy	Medium	Hard
Scale levels	**Node**	Minor	Minor	Moderate
	Local Area Network	Moderate	Significant	Significant
	Enterprise Network	Moderate	Significant	Significant

performed by the threats which directly influence the risk involved. The level of impact can be measured multiple ways, such as

- Magnitude of loss

- Cost of losing data

- Level of damage

- Cost of unavailable service

- Cost of repair

- Time needed for repair

- Cost of gathering the data again

- Indirect costs (trust, reputation)

Scale, detectability, and recoverability assessment

In the impact assessment method, impact depends on two factors, the scale level, and detectability and recoverability levels. The scale level expresses whether the attack only targets and has impact on a particular node or element, and affects the operation of a local part of a system or network, or maybe even the whole system or network is affected by an attack. Therefore three values can be assigned to this property (see also Table 6.2):

- Node: Only the targeted node is affected by the threat; a fault is not propagated to the adjacent nodes/elements

- Local (Area) Network: the threat affects the operation of the local network

- Enterprise Network: the whole network is affected by the threat

The detectability and recoverability property expresses that in case of an attack how hard it is to first detect the attack, avert it, and then recover the system to a state that was present before the attack. These possible values can be assigned to it:

- Easy: Detecting the attack is easy, recovery is quick

- Medium: detecting the attack is moderately difficult or the recovery takes time, possibly requires user intervention, and the exact state cannot be recovered

- Hard: detecting the attack is difficult, distinct algorithms have to be implemented, recovery is not straightforward, requires user intervention, and there is possible data loss involved.

Asset impact and attack intensity assessment

In the ETSI standard the overall impact level is influenced by two factors: the impact on the asset and the attack intensity. An asset is basically en equipment or resource which has value to the organization or the company. The impact of a threat to an asset can be on the following three levels:

- Low: The concerned party is not harmed very strongly; the possible damage is low (value 1)

- Medium: The threat addresses the interests of providers/subscribers and cannot be neglected (value 2)

- High: A basis of business is threatened and severe damage might occur in this context (value 3)

There are also three levels defined for the attack intensity as follows:

- Low: Single instance of attack (value 0)

- Medium: Moderate level of multiple instances (value 1)

- High: Heavy level of multiple instances (value 2)

The two factors considered here are the number of attack instances and the time interval between the attacks. Based on these the impact is calculated by adding the impact and attack intensity values together.

Likelihood assessment

The second part needed for the risk assessment is estimating the likelihood of a threat to be exploited and an that an attack is made to the system.

Two approaches have been considered for likelihood assessment, one based on motivation and difficulty assessment described in [21], the other based on time, expertise, opportunity, and equipment assessment described in ETSI TS 102 165-1 standard [20].

TABLE 6.3

Motivation and difficulty assessment.

Difficulty Motivation	None	Solvable	Strong
Low	Unlikely	Unlikely	Unlikely
Moderate	Likely	Possible	Unlikely
High	Likely	Likely	Unlikely

Motivation and difficulty assessment

Difficulty of performing an attack can be (see also Table 6.3)

- Strong: Security mechanisms that currently may not be defeated because some theoretical elements needed for perpetrating an attack upon them are missing

- Solvable: Security mechanisms that may be countered or have been defeated in a related technology

- None: A precedent for the attack already exists

ETSI likelihood assessment

According to the standard, this is influenced by five factors. Based on these factors a scoring scheme is defined according to Table 6.4.

The scores of these factors are assessed and summarised giving the overall score of the attack likelihood, which is mapped to vulnerability rating according to Table 6.5.

Risk Assessment

The risk score is calculated by multiplying the scores given to the likelihood and impact levels defined in the previous chapter. Based on this score, the risk is determined according to Table 6.6.

The next step is to assign the risks of the threats of the affected security objective types. In the ETSI standards the following security objective types have been defined:

- Interception

  - Eavesdropping: A breach of confidentiality by unauthorised monitoring of communication

- Manipulation

  - Masquerade (spoofing): The pretence of an entity to be a different entity. This may be a basis for other threats like unauthorised access or forgery

TABLE 6.4
Attack likelihood factor scoring.

Factor	Range	Value
Time (1 point per week)	≤ 1 day	0
	≤ 1 week	1
	≤ 1 month	3
	≤ 3 months	13
	≤ 6 month	26
	> 6 month	attack potential is beyond high
Expertise	Layman	0
	Proficient	2
	Expert	5
Knowledge	Public	0
	Restricted	1
	Sensitive	4
	Critical	10
Unnecessary/unlimited access	Unnecessary/ unlimited access	0
	Easy	1
	Moderate	4
	Difficult	12
	None	attack path is not exploitable
Specialised	Standard	0
	Specialised	3
	Bespoke	7

TABLE 6.5
Vulnerability rating.

Attack potential values	Resistant to attacker with attack potential of:
0–2	No rating
3–6	Basic
7–14	Moderate
15–26	High
> 26	Beyond high

TABLE 6.6

Risk assignment table.

Value	Risk	Explanation
1, 2	Minor	No essential assets are concerned, or the attack is unlikely. Threats causing minor risks have no primary need for countermeasures.
3, 4	Major	Threats on relevant assets are likely to occur although their impact is unlikely to be fatal. Major risks should be handled seriously and should be minimized by the appropriate use of countermeasures.
6, 9	Critical	The primary interests of the providers and/or subscribers are threatened and the effort required from a potential attackers to implement the threat(s) is not high. Critical risks should be minimised with highest priority.

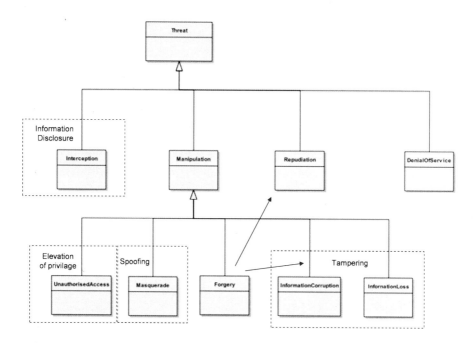

FIGURE 6.13

Assigning STRIDE threat categories to ETSI categories.

- Loss or corruption of information: The integrity of data (transferred) is compromised by unauthorized deletion, insertion, modification, re-ordering, replay, or delay

- Unauthorised access: An entity accesses data in violation of the security policy in force.

- Forgery: An entity fabricates information and claims that such information was received from another entity or sent to another entity.

- Repudiation: An entity involved in a communication exchange subsequently denies the fact.

- Denial of service: An entity fails to perform its function or prevents other entities from performing their functions.

These threat types differ a bit from the STRIDE threat types, but the assignment between them is quite straightforward (see Figure 6.13).

The reasons for having the threat categories for the individual threats is that ETSI defines which of the security objectives are affected by an attack. To wrap up, the threats generated by the STRIDE method based on the DFD are categorized into one of the categories in the STRIDE acronym. These categories can be translated into ETSI threat types.

6.4.3   Safety analysis

6.4.3.1   FMEA/FMECA (Failure Mode Effect Analysis/Failure Mode Effect, and Criticality Analysis)

FMEA - Failure mode effect analysis

FMEA is used to identify potential failure modes, determine their effect on the operations of the product, and identify actions to mitigate failures. FMEA is a bottom-up, inductive analytical method which may be performed at either the functional or component/sub-system level. A successful FMEA activity helps the developer team to identify potential failure modes based on past experience with similar products or processes, and enables them to design those failures out of the system with the minimum of effort and resource expenditure, thereby reducing development time and costs. It is widely used in manufacturing industries in various phases of the product life cycle and is now increasingly finding use in the service industry. FMEA can be applied to hardware as well as software, and it allows a quantitative approach based on known component reliabilities as well as a qualitative one, where no reliability figures are available but experts judge based on their experience.

Figure 6.14 shows the basic steps for performing an FMEA analysis. Based on a system description, the functions of a system are identified. One function is selected and the failure modes of the function are identified. For a failure mode, the effects and causes and the probability are identified. This is repeated

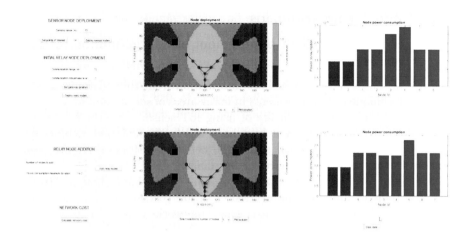

FIGURE 6.15
IoT sensor network deployment tool.

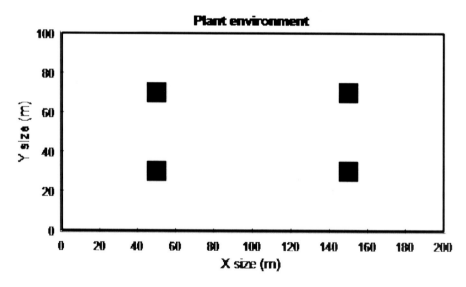

FIGURE 6.16
Plant environment.

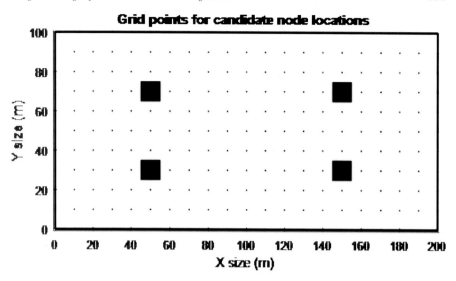

FIGURE 6.17
Available locations for node positioning.

Sensor node deployment

Sensor node locations can be entered directly, or alternatively, points of interest can be entered to get some advice on the sensor node placement. If the points of interest are introduced, a sensing coverage map is represented from the points of interest with the aim of giving advice in the consequent sensor node deployment. For this calculation, nodes' sensing range and obstacle locations are considered. A point in the plant will be covered by a sensor node if, and only if, the two points lie within the sensing range and there is no obstacle between them.

Figure 6.18 shows the target point (or points of interest) input tool. The resulting sensing coverage map is represented in figure 6.19. This coverage map gives a clear idea of the locations for the minimum number of sensor nodes needed to cover all the target points. A location with a sensing coverage redundancy of 4, for instance, means that from that location four target points are covered at the same time. In this example, it is evident that only two sensor nodes would be needed to cover all the points of interest.

Figure 6.20 shows the sensor node input tool, and Figure 6.21 the resulting sensing coverage map. This is the actual sensing coverage map, which is calculated from the sensor node points. It can be seen how all the target points are covered by at least one sensor node.

Alternatively, the sensor node locations can be entered directly. In the sensor node input tool of Figure 6.22, no coverage advice is given. The resulting sensing coverage represented in Figure 6.23 is the same as thats obtained in figure 6.21.

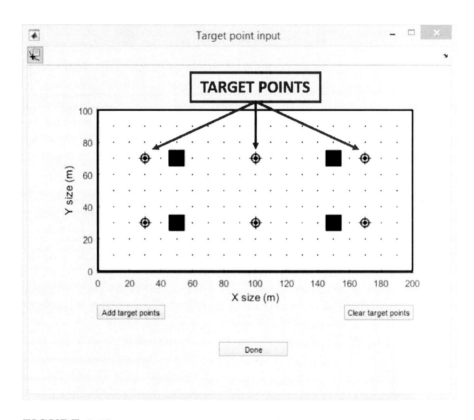

FIGURE 6.18
Target point input tool.

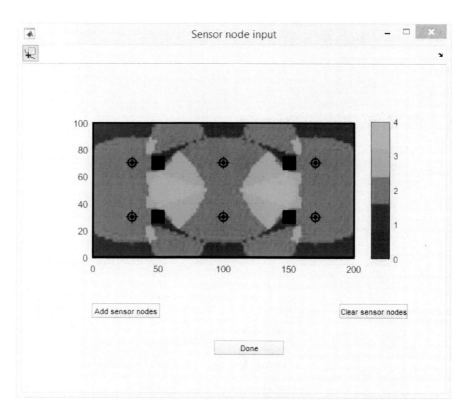

FIGURE 6.19
Sensor node input adviser tool.

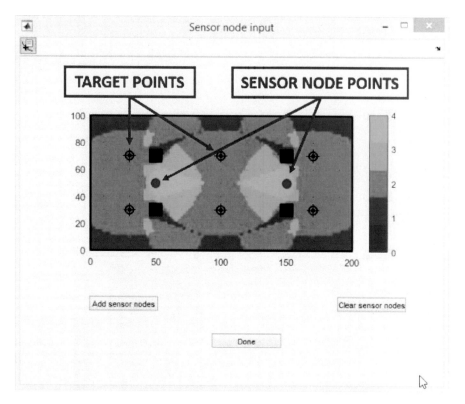

FIGURE 6.20
Sensor node input tool.

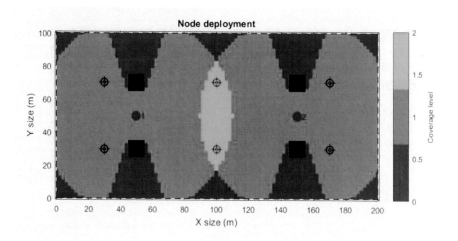

FIGURE 6.21
Sensing coverage map.

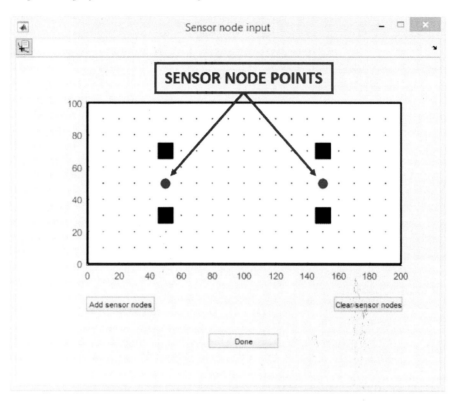

FIGURE 6.22
Sensor node input tool.

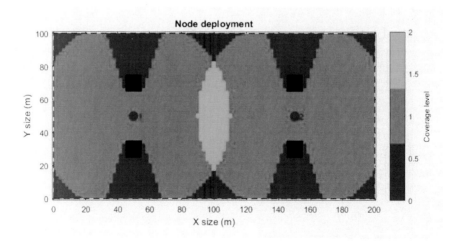

FIGURE 6.23
Sensing coverage map.

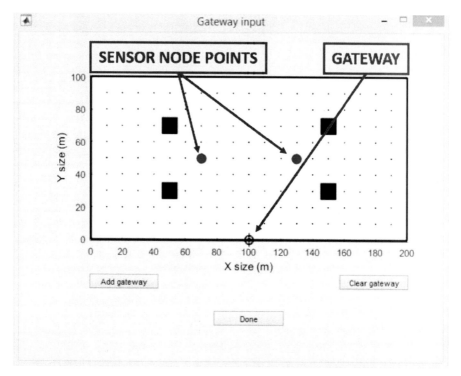

FIGURE 6.24
Gateway input tool

Initial relay node deployment

Once the sensor node locations are determined, relay nodes are added in order to ensure robust connectivity. Every sensor node in the network has to be connected with the gateway directly or by means of relay nodes. To determine whether two nodes are connected or not, an effective distance is calculated (the distance through an obstacle is increased by a given scaling factor). It is considered that two points are connected with each other if the effective distance between them lies inside the communication range.

After determining the gateway position (Figure 6.24), the relay node locations are automatically calculated. The developed algorithm minimises the number of relay nodes by means of sharing relay nodes between different sensor nodes as far as possible. Figure 6.25 shows the resulting scheme for a given example. Sensor node points are represented as red points and relay node points as blue points.

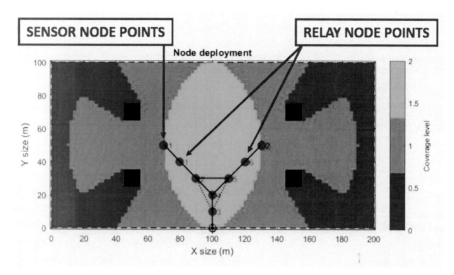

FIGURE 6.25
Relay node deployment

Relay node addition

After the initial node deployment, more relay node points can be added in order to improve network lifetime. This is done by adding more relay nodes near those with higher relaying workload, so that the workload and the lifetime of each node are more evenly distributed.

In Figure 6.26 the initial node deployment (with the minimum number of nodes) is represented. The bar graph of Figure 6.27 represents the resulting nodes' power consumptions red bars correspond to sensor node power consumption, while blue bars correspond to relay node power consumption. Node power consumption are calculated from nodes' transmitting and receiving distances. In Figure 6.28 a redeployment with an extra relay node is represented. From the power consumption represented in Figure 6.29, it can be seen how adding an extra relay node contributes to homogenising the nodes' power consumption.

6.5.1.2 Cost of wireless sensor network

One of the main objectives of the deployment tool is to calculate different deployment solutions in order to determine which solution is the more cost effective. Obviously, a solution with the minimum number of nodes will be the one with less material cost, but, at some time, it may become more expensive in terms of maintenance cost, as it would be necessary to replace some nodes' batteries more often.

A WSN cost tool has been developed to calculate the overall network de-

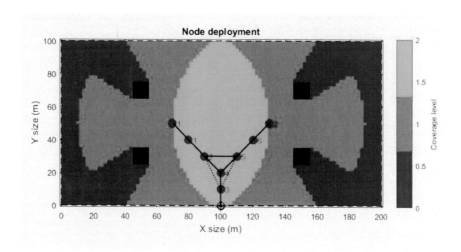

FIGURE 6.26
Initial node deployment (minimum number of nodes).

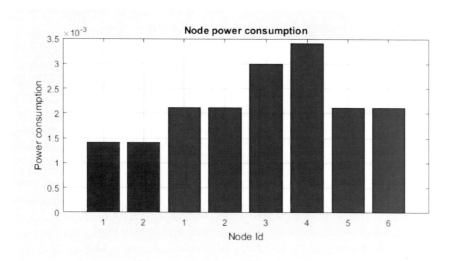

FIGURE 6.27
Initial deployment's power consumption.

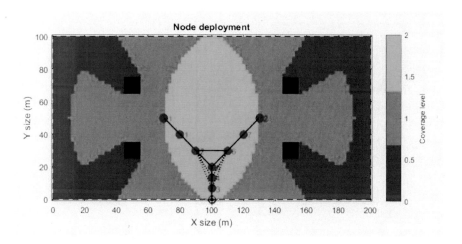

FIGURE 6.28
Node redeployment (one node added).

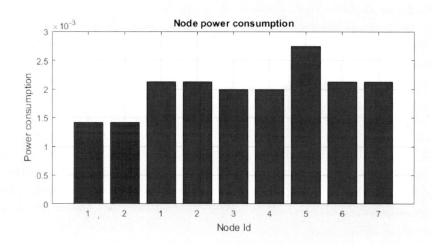

FIGURE 6.29
Redeployment's power consumption.

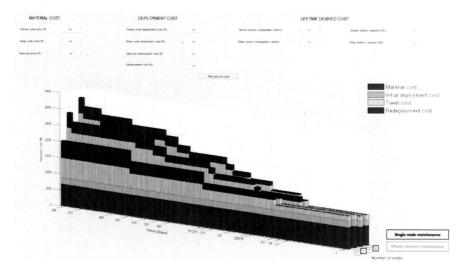

FIGURE 6.30
WSN cost tool.

ployment cost. The contributions considered are material cost, deployment cost, and travel cost. For each solution (that is, solutions with different numbers of nodes) two separate costs have been calculated. In the first case, each node is replaced or maintained at the time its battery runs out, while in the second case, all nodes are replaced at the time the first maintenance service is required.

Figure 6.30 shows the developed cost calculation tool, where specific material and deployment costs can be assigned for sensor and relay nodes (as well as for the gateway). Nodes' power consumption and battery capacity are also required in order to estimate a lifetime derived cost (nodes have to be redeployed as their batteries run out).

For each solution in Figure 6.26 and Figure 6.28 (with six and seven relay nodes), two separate costs are calculated: one for "single node maintenance" and the other for "whole network maintenance." The represented cost is the cumulative sum of the different contributions as indicated in the legend. For the cost parameters considered, it can be seen how the seven relay node solution becomes cheaper in time as nodes in the six node solution have to be replaced more often. Furthermore, it can be seen that "whole network maintenance" becomes cheaper than "single node maintenance" as travel cost has more impact than redeployment cost.

6.5.2   Scenario: Swift deployment and configuration

This scenario reflects the situation where a large number of devices are to be installed in a network where they are to perform different tasks depending or have different configurations depending on the physical location for each device.

The devices are assumed to have identical hardware and identical software preloaded from a factory, workshop, or back office, the only differences between devices are their network Media Access Control address (MAC address) and their Serial Number (S/N). The preloaded software contains the required security measures to allow the device to connect to the Arrowhead Framework core systems and to use a minimal set of services.

The information describing the different tasks and configurations is stored in a network connected storage area accessible from all locations where devices are to be installed and in a format that the devices are able to interpret. The generation and management of this information is not covered in this scenario.

6.5.2.1 Step-by-step general deployment procedure

Initially (after engineering but before deployment) there are the following systems and devices:

- A generic, configurable device, ready to be deployed

- A node in the PlantDescription representing the device that is to be introduced.

- A node in the PlantDescription representing each system on the device is intended to fulfil, each of the system nodes has at least one link to associate it with the device node.

- A configuration in the Configuration store associated with the device node in the PlantDescription. This configuration contains enough information to allow the device to host the desired systems and the NodeId by which each system node is identified in the PlantDescription.

- An identifier that is to be transferred to the device during deployment/-bootstrapping that allows the Configuration system to associate the device with the device-node identifier from the PlantDescription. This may be achieved by using the NodeId from the PlantDescription or by storing a specific identifier, such as a public key from a certificate, in the Configuration system beforehand.

- Each planned service interaction between systems should be stored as service links between the system nodes in the PlantDescription.

- A user interface available during deployment should be able to navigate the PlantDescription and transfer the previously mentioned identifier to or from the device through a local interface.

Deployment procedure:

1. Device is physically connected to the network and turned on.

2. Device connects to the network using DHCP, or a similar standardized technology applicable to the network in question

3. Device looks for the Arrowhead core systems [23] at predefined locations (e.g., on the local network or a cloud service hosted by the device supplier)

4. Core systems authenticate device as factory configured device with basic authorisation

5. The user interface is used to associate the physical device with the device node in the PlantDescription

6. The device registers with mandatory core systems before access is granted to support core systems

7. The device uses the identifier to access the Configuration store and retrieve its configuration.

8. The device configures itself to host systems according to the configuration, and the systems registers with the SystemRegistry.

9. Depending on the implementation of Orchestration used, one of the following interactions is initiated:

   a) The SystemRegistry accesses the PlantDescription providing the association between the SystemId and the NodeId. The PlantDescription system checks for all service links to the associated NodeId and requests SystemId, for all NodeIds connected by service links. Using these provided SystemIds the PlantDescription system creates orchestration rules for all affected systems

   b) The SystemRegistry notifies the Orchestration system with the newly registered SystemId and associated NodeId. The Orchestration system then queries the PlantDescription for system nodes linked to the NodeId by service links. It then uses the provided NodeIds to query the SystemRegistry for associated SystemIds and creates the desired orchestration rules, possibly also using other available information such as QoS or similar

   c) The newly introduced system requests an orchestration from the Orchestration system, providing either SystemId or its NodeId. The Orchestration system in turn accesses the PlantDescription (if necessary after getting the NodeId from the SystemRegistry) which responds with the appropriate service links. Using the service links, and other available information, the Orchestration system provides the requested orchestration back to the newly introduced system.

At the end of the procedure, the operators and engineers should be informed the procedure has been successful or if some part of it failed. This can be done through an event or subscription mechanism, where involved user interfaces are notified depending on the result. For example, a deployment failure could first be sent to the user interface that was used in the deployment process, and only escalated to other users if the user that tried to deploy the device is unable to solve the issue.

6.5.3 Scenario: PLC device monitoring

On PLC controlled industrial equipment, services are likely to be provided not by the device itself for obvious reasons related to security and production contraints, but by a higher-level network node to which the PLC device is connected. OPC/OPC UA is a widely accepted protocol for interfacing with PLC controllers and was used in the Arrowhead project to develop a tool set used to support automation system life cycle (i.e., design, deployment, operation monitoring, and maintenance). The OPC UA server is the interface to PLC devices and is used to produce/consume services that cannot be deployed at device levels.

Two main scenarios of the automation life cycle can be identified during or after the commissioning phase:

- System configuration

- System monitoring

The configuration of the PLC control system requires downloading of PLC code to the target controller. For safety, security, and practical reasons, this is typically achieved through specific and/or proprietary software using a direct connection to the PLC device, and not via web service provided or consumed by the device itself. The provision of web service is implemented at a higher level of the system architecture.

6.5.3.1 Configuration management

For PLC devices which code is generated using the CCE Mapper tools [23], the configuration data for a PLC consists of

- PLC control code

- Component to FB (Function Blocks) mapping

- Function block to I/O mapping

- Variable list (that needs to be monitored)

- PLC memory addresses/OPC UA variable mapping

- OPC UA server IP (for configuring OPC UA client)

- Service provider server IP (for configuring OPC UA client)

Configuration data generated by the Control Auto-generation tool can be versioned and stored on the application server for later retrieval by the mapper tool. The configuration data is used by the mapper tool to configure the PLC device (i.e., download of the control code) and configure the OPC UA server and OPC UA client. Once the OPC UA client is configured, it connects to both the OPC UA server and application server. The system is then ready to collect data.

The PLC generated data (e.g., state change, fault) are collected by the OPC UA server. The OPC UA client polls or is notified of PLC events and pushes the data to the application server, where the data is stored. End user applications or higher level application servers can consume and process the data.

Example of services provided by the application server layer and related to this operation and maintenance phase related scenarios are listed below:

- StoreConfig: allows the deployment tool to store a new OPC UA server configuration (i.e., OPC UA variable to PLC memory address mapping)

- GetConfig (SysId, Version): allows retrieval of a specific version of configuration data for a specific system

- GetSystemList: returns a list of systems for which configuration data is stored

- GetLog(SysId, VarTyp, start$_{date}$, end$_{date}$): Gets log of variable change for a given system, for a given variable type and time interval

6.5.4   Scenario: Replacement of device

For a device containing its own configuration, program code, or other local customisation, the procedure for replacement becomes much more complex than that of simply replacing a generic device that always operates in the same mode. If the component that has failed is not capable of providing a copy of its data after the failure, it is very important that up to date documentation and backup of data is readily available.

6.5.4.1   Common current procedure

The current procedure is that all forms of configurations, code and customisations are documented and stored either electronically or on paper in a repository or archive, managed by the site owner, system supplier or a system integrator. In the event of replacement of such a device, a new device is acquired and the party responsible for maintaining the archive or repository configures the device before it is tested and installed at the site where the functionality is once again verified.

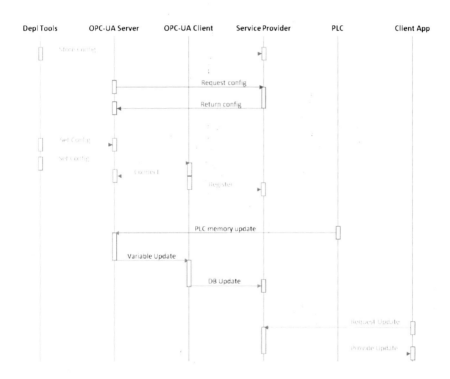

FIGURE 6.31

Retrieval of OPC UA client server connection data to monitor PLC device states.

6.5.4.2   Recommended procedure for Arrowhead

As all Arrowhead devices are expected to be interoperable, it is recommended that they can support some method of retrieving all device-specific data and that this data can be used to easily deploy a replacement device.

The proposed method for this is for the relevant data to be stored at a network-accessible location, with backups, redundancies, and security measures as deemed necessary. As a replacement device is connected and authorised, the stored data can then be accessed directly and assigned to a new device through a few services, requiring less engineering effort.

This method of replacement is intended to be enabled by the Configuration system and PlantDesrciption system. Further support is given through the deployment procedure that is to be used to give a newly connected device its initial network access and authorisation so that it is able to find the systems providing Configuration services.

The Configuration services are intended to provide the functionality for storing and retrieving the device-specific data, for initial configuration of systems as well as reconfiguration and replacement of devices.

The PlantDescription services are intended to give a basic common understanding of the layout of the plant or site, providing possibilities for actors with different interests and viewpoints to access their view of the same dataset. In the case of device replacement, this is useful for the technician replacing the device to assign which position the new device is in, e.g., which old device it is meant to replace.

6.5.4.3   Specific cases - Device configuration

In some cases and for some types of devices (e.g., many PLC controllers), the services required to achieve device configuration cannot be accessed by the device itself (e.g. no direct access to the network of services, limited resources, security/safety reasons, etc.). In this case the services required to achieve device configuration are consumed by an external configuration device (e.g., Laptop, machine HMI) physically connected to the PLC.

6.5.4.4   Specific cases - Device hardware versions/types

A given PLC control device might be replaced with a device from another vendor. A typical scenario is update of control systems on legacy production equipment. From an Arrowhead/Engineering scenario perspective, this implies that a given device configuration may exist in the device configuration database in several formats. In such a case, the Configuration service would therefore have to provide the ability to select configuration format as well as configuration version.

6.5.4.5 Specific cases - Device configuration formats

Device configuration might exist in various formats throughout the device life cycle. During virtual design, and virtual validation, device configuration may be generically described (e.g., XML-based description of PLC control logic). The generic description will at some point be translated into device/vendor specific format (e.g., Siemens Step 7 files) and compiled during installation on the target hardware. The Configuration service could therefore provide the ability to retrieve a given device configuration in several formats, depending on the end user requirements (e.g., update of configuration on the device itself, or device configuration design modification). Possibly services related to checking consistency between the same device configuration in various formats would also be necessary in order to support a scenario such as the one described in Section 6.5.5. The need for device configuration format translation could also be investigated.

6.5.5 Scenario: Device configuration upload

This scenario is based on a common but not always suitable practice typically encountered in industry, which results in the device configuration being tweaked by shop floor technicians and engineers either because the initial device design configuration was flawed or because the machines physical configuration was changed and therefore the control configuration needs to be changed accordingly. In such case the software/configuration installed on the physical device may become the gold standard and should be stored in the configuration database as a new update/version of the previous configuration. The Arrowhead configuration service should therefore provide the ability to allow a device (or any device configuration system) to upload a new or update a version of the configuration in the configuration database.

Bibliography

[1] R. Yang, *Process Plant Lifecycle Information Management*. Author-House, 2009. [Online]. Available: https://books.google.se/books?id=zmgzdjf1GR0C

[2] A. J. Braaksma, W. W. Klingenberg, and P. P. van Exel, "A review of the use of asset information standards for collaboration in the process industry," *Computers in Industry*, vol. 62, no. 3, pp. 337 – 350, 2011. [Online]. Available: http://www.sciencedirect.com/science/article/pii/S0166361510001387

[3] *IEC 62424 Representation of process control engineering - Requests in P&I diagrams and data exchange between P&ID tools and PCE-CAE tools*, International Electrotechnical Commission Std.

[4] R. Drath, A. Luder, J. Peschke, and L. Hundt, "Automationml - the glue for seamless automation engineering," in *2008 IEEE International Conference on Emerging Technologies and Factory Automation*, Sept 2008, pp. 616–623.

[5] *MIMOSA OSA-EAI - Open System Architecture for Enterprise Application Integration*, MIMOSA OSA-EAI Std. [Online]. Available: http://www.mimosa.org/mimosa-osa-eai

[6] W. Mahnke, S.-H. Leitner, and M. Damm, *OPC Unified Architecture*. Springer, 2009.

[7] V. Vyatkin, *IEC 61499 Function Blocks for Embedded and Distributed Control Systems Design*, third edition ed. International Society of Automation, 2016.

[8] ISO/IEC 81346-1:2009, Industrial systems, installations and equipment and industrial products - Structuring principles and reference designations, ISO/IEC Std.

[9] *ISO 15926 Industrial automation systems and integration - Integration of life-cycle data for process plants including oil and gas production facilities*, International Organization for Standardization Std.

[10] *fastAPI - Standard för fastighetskommunikation*, SABO Std. [Online]. Available: http://www.fastapi.se/

[11] *IEC 61850: Power Utility Automation*, International Electrotechnical Commission Std.

[12] M. Schleipen. (2014) Open standards for Industry 4.0 - Tools and offer around AutomationML and OPC UA. [Online]. Available: http://www.iosb.fraunhofer.de/servlet/is/46944/AutomationML_en.pdf

[13] A. T. Johnston, "OpenO&M and ISO 15926 Collaborative Deployment," October 2009. [Online]. Available: http://www.mimosa.org/presentations/openom-and-iso-15926-collaborative-deployment

[14] P. Adolphs, H. Bedenbender, D. Dirzus, M. Ehlich, U. Epple, M. Hankel, R. Heidel, M. Hoffmeister, H. Huhle, B. Kürcher, H. Koziolek, R. Pichler, S. Pollmeier, A. Walter, B. Waser, and M. Wollschlaeger, "Status Report - Reference Architecture Model Industrie 4.0 (RAMI4.0)," VDI - Verein Deutscher Ingenieure e.V. and ZVEI - German Electrical and Electronic Manufacturers' Association, Tech. Rep., July 2015. [Online]. Available: http://www.zvei.org/Downloads/Automation/

5305PublikationGMAStatusReportZVEIReferenceArchitectureModel.
pdf

[15] H. Kagermann, W. Wahlster, and J. Helbig, "Recommendations for implementing the strategic initiative industrie 4.0," acatech - National Academy of Science and Engineering, Tech. Rep., 2013.

[16] S. Plosz, M. Tauber, and P. Varga, "Information Assurance System in the Arrowhead Project," *ERCIM News*, vol. 97, p. 29, April 2014, iSSN 0926-4981.

[17] *ISO/IEC 27005 , Information technology - Security techniques - Information security risk management*, International Organization for Standardization Std., 2011.

[18] M. Howard and D. E. Leblanc, *Writing Secure Code*, 2nd ed. Redmond, WA, USA: Microsoft Press, 2002.

[19] S. Hernan, S. Lambert, T. Ostwald, and A. Shostack, "Uncover security design flaws using the STRIDE approach," *MSDN Magazine*, Nov. 2006. [Online]. Available: http://msdn.microsoft.com/en-us/magazine/cc163519.aspx

[20] *Telecommunications and internet protocol harmonization over networks (TIPHON) release 4; protocol framework definition; methods and protocols for security; part 1: Threat analysis*, ETSI Technical Specification, 2003.

[21] M. Barbeau, "Wireless security in the home and office environment," Technical Reports, Carlton University, Tech. Rep., 2010.

[22] C. Schmittner, T. Gruber, P. Puschner, and E. Schoitsch, *Computer Safety, Reliability, and Security: 33rd International Conference, SAFECOMP 2014, Florence, Italy, September 10-12, 2014. Proceedings.* Cham: Springer International Publishing, 2014, ch. Security Application of Failure Mode and Effect Analysis (FMEA), pp. 310–325. [Online]. Available: http://dx.doi.org/10.1007/978-3-319-10506-2_21

[23] X. Kong, B. Ahmad, R. Harrison, A. Jain, Y. Park, and L. J. Lee, "Realising the open virtual commissioning of modular automation systems." in *Proceedings 7th International Conference on Digital Enterprise Technology*, Athens, Greece, 2011, pp. 1–9.

7

Application system design - energy optimisation

Michele Albano
ISEP, Polytechnic Institute of Porto

Arne Skou
Aalborg University

Luis L Ferreira
ISEP, Polytechnic Institute of Porto

Thibaut Le Guilly
Aalborg University

Per D Pedersen
Neogrid

Torben Bach Pedersen
Aalborg University

Petur Olsen
Aalborg University

Laurynas Šikšnys
Aalborg University

Radislav Smid
Czech Technical University

Petr Stluka
Honeywell

Claude Le Pape
Schneider Electric

Chloé Desdouits

Schneider Electric

Rodrigo Castiñeira

Indra

Rafael Socorro

Acciona

Inge Isasa

Orona

Jani Jokinen

Tampere University of Technology

Lorenzo Manero

IK4-IKERLAN

Aitor Milo

IK4-IKERLAN

Javier Monge

Indra

Anatolijs Zabasta

Riga Technical University

Kaspars Kondratjevs

Riga Technical University

Nadezhda Kunicina

Riga Technical University

CONTENTS

7.1 Introduction

In this chapter, we present a number of applications of the Arrowhead Framework with special attention to services related to awareness and optimisation of energy consumption. First, we present the notion of FlexOffers as a general mechanism for describing energy flexibility. FlexOffers can be aggregated into larger flexibility units to be used as an Arrowhead service in the virtual market of energy [1]. This is followed by two examples on how to exploit such a flexibility service in the energy management of heat pumps and a campus building. Then we present two examples on how to exploit renewable energy to provide elevator services. Next, two examples of context aware services are described – smart lighting and smart car heating, and finally it is described how the Arrowhead Framework can play a role in the optimisation of municipal service systems. In the final section, we indicate future work.

7.2   Market as an energy optimising method

This section describes the virtual market of energy [2], which is one of the demonstration platforms using the Arrowhead Framework. It builds upon the Danish national ForskEL project "TotalFlex[1]."

7.2.1   Energy flexibility concept

The virtual market of energy is based on the notion of flexibility, i.e., the ability to postpone or bring forward the operation of a component in time, or increase or decrease its energy consumption for a given period. A component enabling flexible control of its operation is called a flexible resource (FR). The owner of an FR is referred to as a prosumer. The flexibility market is part of the virtual market of energy demonstration platform, and it is the electronic market where flexibility is traded.

7.2.1.1   FlexOffers

To be able to exchange information about energy flexibility among different actors of the flexibility market, there is a need for a common representation of flexible load. The European project MIRABEL proposed a format to encode this information, called FlexOffer.

A visual representation of a simple FlexOffer is shown in Figure 7.1. Each bar in the graph corresponds to a time slice of energy consumption, with the lower part representing the minimum amount of energy that a flexible resource needs to provide its service, and the upper part an interval in which it can adjust its consumption, while still satisfying functional constraints (e.g. comfort temperature). This corresponds to energy flexibility. Another type of flexibility is time flexibility as shown in Figure 7.1. Time flexibility is provided when an energy load can be shifted within a time interval, defined by an earliest start time at which the FR can start its consumption, and a latest end at which it should be done. When created, a FlexOffer is assigned a baseline schedule that corresponds to the consumption pattern that the associated flexible resource prefers and will follow. Updated schedules can be assigned to the FlexOffer to modify the consumption behaviour of the flexible resource, utilizing its provided flexibility.

7.2.1.2   Aggregation

FlexOffers from individual flexible resources (e.g., heat pumps, electric vehicles) most often do not represent large flexible loads. Thus, a single such FlexOffer has low impact and is of little interest for balancing demand and supply loads on the grid, where required balancing capacities are much higher.

---

[1] www.totalflex.dk

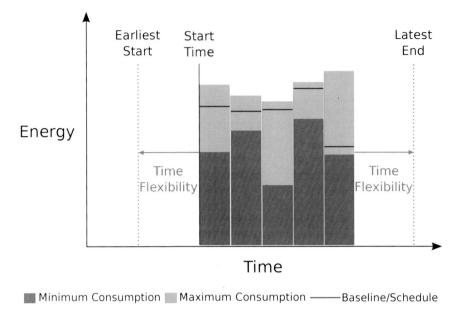

FIGURE 7.1
Example of a FlexOffer.

At the same time, optimising energy loads based on large numbers of Flex-Offers is a computationally hard problem, which requires dealing with many decision variables and constraints originating from many FlexOffers. A common solution to all these problems is FlexOffer aggregation, elaborated next.

As seen in Figure 7.2, individual FlexOffers can be aggregated into so-called aggregated FlexOffers. Aggregated FlexOffers specify larger energy amounts with associated aggregated flexibilities, while taking the same representation, and being managed in the same way, as individual FlexOffers. Note that aggregation can be performed multiple times, meaning that an aggregated FlexOffer can, potentially, be composed of (smaller) aggregated FlexOffers.

During energy optimisation (FlexOffer scheduling), an aggregated FlexOffer is assigned a schedule, which, by respecting all inherent aggregated FlexOffer constraints, specifies an exact start time and energy amounts of an aggregated load to be assigned to a number of underlying flexible resources. Such a schedule needs to be disaggregated to a number of schedules for each individual FlexOffer it is composed of, and so this operation is denoted FlexOffer disaggregation. While disaggregating schedules, as demonstrated in Figure 7.2, it is important to respect all (time and energy) constraints of every FlexOffer as well as to ensure that the energy amounts at every time slice are equal before and after the disaggregation. Otherwise, significant losses associated with imbalances will be encountered.

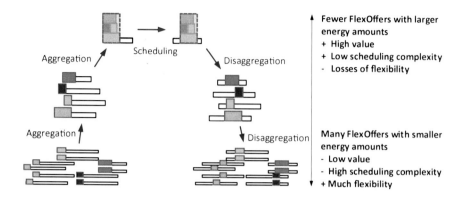

FIGURE 7.2
Workflow of FlexOffer aggregation and disaggregation.

In general, the flexibility of aggregated FlexOffers tends to be lower than the joint flexibility of the FlexOffers that compose them. This reduction in flexibility is, however, unavoidable in order to reduce FlexOffer scheduling complexity and to increase their value (e.g., on the flexibility market). Note also that aggregating flexible loads specified as FlexOffers is a complex task. To optimise aggregation and disaggregation, FlexOffers need to be grouped together based on similarity of their flexibility patterns. Algorithms for FlexOffer aggregation and disaggregation have been provided [3].

7.2.2   Flexibility market framework

The *vision* is to show a working implementation of the flexibility market and the benefits of using the Arrowhead framework. The objective is to demonstrate the flow of FlexOffers from creation to execution, using different kinds of flexibility operating in live mode during different phases:

(i) Creation

(ii) Aggregation

(iii) Scheduling

(iv) Disaggregation

(v) Execution

7.2.2.1   Main actors

The main actors in a Flexoffer market are listed below while their relationships are visualised in Figure 7.3.

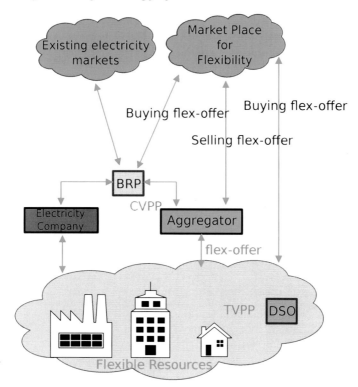

FIGURE 7.3
Overview of the main actors.

Prosumer Owns the flexible resource that delivers the flexibility.

DSO[2] Grid operator and buyer on the flexibility market.

Aggregator Collects and aggregates the FlexOffers from Prosumer and acts as a seller on the flexibility market.

Flexibility Market The place where flexibility is traded.

BRP[3] Balancing responsible party, secures balance in an area and has access to existing energy markets and is a buyer on the flexibility market.

Electricity Company Takes care of selling and buying energy for the Prosumer.

---

[2]DSO: Distribution System Operator has the grid responsibility for medium to low voltage power distribution in a geographical area.

[3]BRP: Balance Responsible Party has been delegated a balance responsibility from the TSO (Transmission System Operator, i.e. Energinet in DK) for part of its area

In the virtual market of energy, the commodity traded is not energy but possibilities of deviation according to a reference consumption plan. The flexible resources are separated from the conventional non-flexible devices in a home. This means they are metered and settled separately. This is also in line with the ideas in the proposed Danish "Market model 2.0"[4]. In the framework, the flexible resources are connected to an aggregator, which also has a role as electricity company. The aggregator has an agreement with a BRP to purchase its energy and cover eventual imbalances.

FlexOffers are generated locally at the flexible resource. They might as well be generated at the Aggregator after collection of measurement data.

To overview and plan the operation of the distribution grid, the DSO is using the IT tool TVPP[5] and, similarly, the BRP and Aggregator are using the tool CVPP[6] to monitor and control the collection of its Flexible Resources as a single power plant. In the flexibility market, the aggregator submits selling FlexOffers with price information, and the BRP and DSO submit buying FlexOffers with price information.

7.2.2.2   Market place clearing process

The basic operations of the key actors are shown in the Figure 7.4. As can be seen in the sequence diagram, FlexOffers are now the method used both on the buyer (DSO and BRP) side and the seller. The FlexOffer includes the willingness to pay and is input to the market place. After FlexOffers are scheduled the Aggregator can update the BRP about the "new" plan for its portfolio, which will then be the basis for actual purchase and selling of energy. The aggregator also disaggregates the Flexoffer and send operation schedules back to the flexible resources. When the FlexOffers are activated, the aggregator measure the operation of the flexible resource to be sure it has operated as planned.

7.2.3   Architecture overview

The flexibility framework is implemented following an architecture articulated into a number of software components that enable information exchange between the actors of the scenario. The main functionalities provided are flexibility forecast, aggregation of FlexOffers, and control of flexible resources.

The Arrowhead Framework acts as an enabler to interconnect different application domains. In this context, the virtual market of energy is a central point in the multi-application domain vision, as it extends to the energy plane the interconnection provided by the Arrowhead Framework for the data plane.

---

[4]Energinet: Market model 2.0 `http://energinet.dk/DA/El/Engrosmarked/Ny\%20-markeds-model/Sider/default.aspx`

[5]TVPP: Technical Virtual Power Plant: IT tool to monitor and forecast grid load.

[6]CVPP: Commercial Virtual Power Plant: IT tool to monitor and control flexible resources such as power plants.

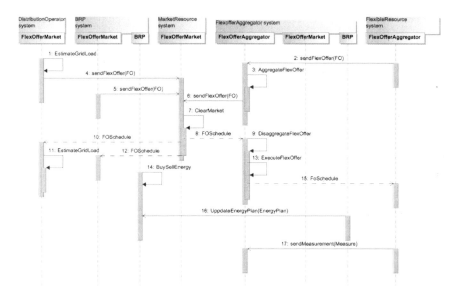

FIGURE 7.4
Basic interactions between key services in the FlexOffer market.

The components of this application scenario provide for three main functionalities:

- generating of FlexOffers on behalf of flexible resources, and validating them;

- controlling the flexible resources energy consumption to enforce energy consumption constraints set by the energy market;

- validating the consumption of flexible resources with respect to their assigned schedules.

Figure 7.5 illustrates the different system components as well as their interconnections. The three main components of the system contain the business logic for the previously introduced actors and their systems and services, namely, FlexibleResource (FR), AggregatorResource (AR) and MarketResource (MR). The Arrowhead core system components provide services to facilitate consumption of the services produced by the main actors.

7.2.3.1 FlexibleResource

The FlexibleResource system enables the management of a flexible energy load on the prosumer side. The functionalities handled by FlexibleResource are:

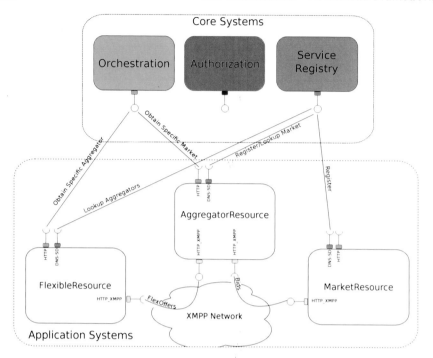

FIGURE 7.5

Software architecture of the Virtual Market of Energy

- FlexOffer generation, which consists in forecasting the flexibility of the underlying system and estimating a safe interval in which it can be operated;

- Consumption management, based on the schedules received back on the proposed FlexOffers.

The architecture is also able to adapt the FlexOffer concept to any platform by developing interfaces towards other elements:

(i) local and remote FlexibleResource

(ii) with the controlled devices' hardware

(iii) with other needed devices (e.g., a remote power meter) through a network

(iv) with external services (e.g., to obtain weather forecasts).

These interfaces can be Arrowhead-compliant or custom, and will not be further described here. It is assumed that, for any type of distributed energy resource, a profile can be defined, the latter containing the current configuration of the distributed energy resource.

It is worth noting that the FlexibleResource, in the software architecture, is only defined as an interface towards an AggregatorResource, and does not enforce any further architectural constraints. This is important due to the large variety of FlexibleResources that can co-exist, which would make it difficult to define a generic architecture for it. This profile could for example contain the comfort settings for a heat pump or the maximum temperature variation for an aluminum smelter.

7.2.3.2 AggregatorResource

The AggregatorResource system implements the functionalities offered by an Aggregator. It acts as a service provider towards FlexibleResources, enabling them to submit generated FlexOffers, informing them on their execution, and sending back consumption schedules when flexible resources baseline consumption patterns are modified.

The AggregatorResource's workflow consists of receiving FlexOffers from multiple FRs, aggregating together the FlexOffers into larger macro FlexOffers, and placing them on the virtual market of energy by providing them to the MarketResource. Afterwards, the AggregatorResource receives a response from the MarketResource, disaggregates the response and sends a consumption schedule to each FlexibleResource.

Several types of AggregatorResources might exist with varying degrees of specialisation for specific scenarios. For example, some AggregatorResources can be focused on the control of electric motors, while others can be more adequate for the control of heating systems. The correct AggregatorResource for the FlexOffers at hand is determined by the Arrowhead Framework Orchestration system at the flexible resource.

7.2.3.3 MarketResource

The MarketResource is the system that encapsulates the energy markets and provides an abstraction layer to ease the access to multiple markets at the same time. The MarketResource receives a (macro) FlexOffer from the AggregatorResource and puts it on one or more energy markets to buy, sell, or store energy, and responds to the AggregatorResource the results of the transaction.

7.2.3.4 Communication Infrastructure

The communication between the FlexibleResource, the AggregatorResource, and the MarketResource is implemented using *HTTP* over *XMPP* protocol [4], and exploits the Arrowhead Framework to establish connection. The main advantage of XMPP is its capabilities to support the publish/subscribe communication paradigm, which provides an asynchronous and highly scalable many-to-many communication model [5]. The resulting decoupling between publishers and subscribers, in time, space, and synchronisation, simplifies the

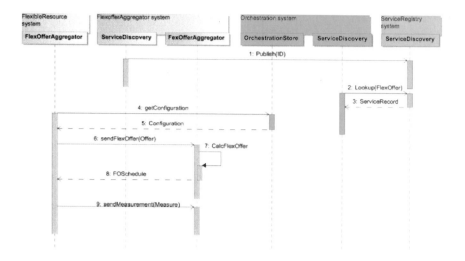

FIGURE 7.6

Sequence diagram for a common virtual market of energy interaction.

implementation of its associated software. Additionally, XMPP is also in process of being standardised as a protocol for the control of demand response applications for OpenADR [6] and on the ISO/IEC/IEEE 21451-1-4 [7] standards.

The interaction with Arrowhead core systems can obey other interaction paradigms. For example, interaction with the ServiceRegistry system. Figure 7.6 represents the typical interactions involved in the provision of FlexOffers from the FlexibleResources to the AggregatorResources, which features a first phase related to the service registration and discovery, and later the provision to the AggregatorResources of information regarding the flexibility in energy consumption contained in the FlexOffer.

As far as security is concerned, the rationale is that the Authorization system is managed by the electric distribution system operator, with whom the prosumer is supposed to have signed a contract to benefit from FlexOffers. Authentication is implemented by a combination of public key Infrastructure (PKI) and X.509 certificates, which the AggregatorResource and the FlexibleResource must obtain from the Authorisation system using the REST [8] protocol. Later on, communications between the actors are encrypted by using XMPP [4] over TransportLayerSecurity (TLS) [9]. Other characteristics and non-functional requirements, such as Quality of Service, can be supported by means of the automation support core services of the Arrowhead Framework [10].

7.3 Optimisations based on a virtual market of energy

Here is described two implementations where the Flexoffer market is used for energy optimisation. The first one concerns control of heat pumps for residential heating. The second one addresses the energy control of a university building.

7.3.1 Heat pumps

The first pilot consists of an individual control of heat pumps installed in occupied residential houses. Each household is provided access to a web application through a FlexibleResource service, enabling setting of comfort temperatures, as shown in Figure 7.7.

FIGURE 7.7
Web application enabling configuration of heat pump set point

The optimisation idea is that a heat pump can be controlled in a flexible manner both in time and energy consumption, while ensuring user constraints such as comfortable temperature interval of the house, for example [11]. The process used in the first pilot focused on intrahour operation and included the following steps:

- Create a model which can predict the energy demand of the house;

- Calculate day-ahead the cheapest energy plan to secure comfort using the spot price;

- Every 15 minutes issue a FlexOffer describing the options for decreasing/increasing power consumption;

- Wait for an eventual FlexOffer schedule.

The first pilot is individual control of heat pumps installed in occupied residential houses. The house has an app where the desired comfort temperature and accepted interval can be set, as shown in Figure 7.7. The concept is that the operation of the heat pump can be shifted in time without loss of comfort and this is its flexibility. The data used for modelling are historical data for

- Delivered heat in the house;

- Used energy for hot water;

- Indoor temperature;

- Power consumption of the heat pump;

- Weather data.

   The data used for forecasting are

   - Created model;

   - Weather forecast;

   - User comfort criteria.

The benefit of the Arrowhead Framework for this pilot is to facilitate the interconnection between heat pumps and aggregator. Heat pumps and aggregators will register is services and systems to the ServiceRegistry and the SystemRegistry of the local clouds involved. The Orchestration system of each local cloud can then distribute orchestration rules to heat pumps thus enabling them to consume the services of an Aggregator. If the Aggregator not is located in the same local cloud and the heat pump the Gatekeeper support system enables discovery, authorisation and orchestration between local clouds. In this way a flexible energy market covering a large geographical area can be generated.

7.3.2 Campus building

The other Flexoffer implementation is in the new building of the University Centre for Energy Efficient Buildings (UCEEB), which is located in Buštěhrad, Czech Republic. The UCEEB building serves as a complex experimental platform for all research fields related to the area of energy efficient buildings. It integrates a variety of spaces, including open space office, smaller office rooms, large halls, and laboratories. It is also an experimental facility with multiple local energy sources integrated into respective distribution grids: one for electricity, one for heat and one for cooling. Individual energy resources include two photovoltaic fields, a combined heat and power (CHP) unit producing heat and electricity, two charging stations for electric cars, storage units for heat and cold, two gas boilers, and two chillers.

Prior to the implementation of the FlexOffer concept, several application scenarios were analysed to determine how the building systems could operate in combination with FlexOffers.

- **Adjusting the HVAC system set points** based on FlexOffers, ensuring the comfort temperature while maximising the economic benefits from active participation in the flexibility market

- Enable generation of FlexOffers by **manipulation with the temperature of hot and cold water**

- Using strategies such as **dynamic pre-cooling or pre-heating** to increase the flexibility of HVAC system during the periods of peak energy consumption.

Several rooms with associated electric heaters and fan-coil units were selected for the implementation in a way that allows experimentation during both hot and cold seasons. An important part of the pilot is a software application that is connected with the Building Management System (BMS) and all archived data. It is used by the building operator to control the process of generation of FlexOffers. This human supervision is important because specific adjustments of the HVAC system operation can potentially lead to uncomfortable situations for the occupants. The building operator is able to balance between comfort level and economic benefits. When the operator interacts with the application, the following functions are provided:

Precool specifies a time interval in which pre-cooling of the building is allowed;

Duty Cycle enforces a specified cycling pattern on the functioning of the HVAC system, which can be used to provide flexibility;

Full Flexibility leaves full control of the system to the FlexOffer framework for a given time period, thus ignoring any comfort setting;

Adjust Set Points allows to manipulate with room temperatures in the building, but in this case the flexibility is specified directly by the operator.

The implementation allowed the assessment of potential economic benefits as well as application constraints.

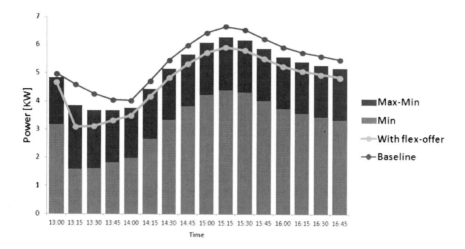

FIGURE 7.8
Load shedding example.

- The FlexOffer concept has certain similarities with applications of the Automated Demand Response (ADR) in commercial buildings. One possible difference is that today's ADR always triggers a firmly defined load shedding strategy, while FlexOffers are assuming more degrees of freedom in building operation. For this reason it seems important to have the possibility of a human override of the overall process of generating FlexOffers.

- Since any flexibility in the operation of HVAC system needs to be seen as a compromise between the energy savings and overall comfort conditions in the interior spaces, it is desirable to implement only relatively short alterations of the default control strategy. This applies primarily to the strategies related to duty cycling and set point adjustments, which both should not span longer time intervals. An example of a load shedding strategy implemented as a FlexOffer, lasting over 4 hours, is illustrated in Figure 7.8, which shows a systematic reduction compared to the estimated baseline.

- The economic benefits of FlexOffers need to be further studied because some reduction of occupants' comfort may cause a so-called rebound effect – known from ADR projects – where the occupants may temporarily increase their heating/cooling demands just after the completion of a given FlexOffer event to compensate for the previous discomfort. This can increase energy consumption and operational costs and to some extent reduce the overall economic benefits. It is important for the stakeholders implementing the FlexOffer concept through energy market mechanisms to incentivise building owners enough to be benefiting from their active participation.

7.4 Energy optimisation on lifts

7.4.1 Multisource elevator energy optimisation and control

Reducing energy demand and cost is a crucial need and a major topic for both the scientific community and industry. Even though details vary from one application domain to another, common features are

- Measuring energy consumption (or production), with an appropriate precision, in time and space, to understand how much energy is consumed or produced by different sub-systems or sub-processes.

- Building a model of energy consumption (or production), explaining how the consumption or production varies with the context. This model serves as a basis for (i) the identification of the flexibility that could be exploited to reduce energy costs and (ii) the actual exploitation of this flexibility from prediction, planning, and control.

- Predicting either the forthcoming energy consumption (or production) directly, or the values of the factors that will influence it. For example, one could predict the future energy consumption of an elevator, either based on historical energy data, or based on historical elevator usage coupled with an elevator energy model.

- Planning and scheduling the consuming (or producing) activities, in order to minimise (maximise) either the consumption (production) itself, its expected cost (revenue), or its environmental impact. In a demand-response context, this includes the identification of possible adaptations of energy consumption (or production) to potential global peak reduction requirements.

- Controlling the relevant sub-systems in accordance with the plan, but enabling the deviations required due to differences between the prediction and the actual execution, occurrences of unexpected events, or the actual implementation of a proposed peak reduction action.

As part of the Arrowhead project, we have designed and implemented a common architecture for three energy cost optimisation applications. This includes

- The use of a common object model to relate the different system components.

- A common infrastructure based on Schneider Electric wireless sensors and on the CEA LINC middleware for energy measurements [12].

- Common services as a basis for the implementation of the previously introduced features: an Energy Tariff service to manage the data enabling to relate energy consumption and cost

  - a DemandPrediction service (applied to both electricity and water demand) to predict resource consumption based on historical data

  - a generic Optimisation service, enabling the resolution of complex optimization models on a remote server.

- The generalised use of web services to link the various software components composing each application (see Figure 7.9).

These common elements are used in three energy cost optimization pilots in manufacturing, elevator/hoisting, and water distribution, enabling savings of the energy cost between 5% and 65%, depending on the application and context. More details on the elevator case are provided in the remainder of this section.

New generations of elevators are equipped with energy storage (i.e., batteries and/or super-capacitors) to allow a minimum of autonomy in case of general power failure. This autonomy is crucial for safety (e.g., to evacuate people with reduced mobility) but energy storage may also offer flexibility in power management. It could be used to decrease the energy bill and environmental impact. For instance, the energy produced when the elevator brakes may be stored instead of evacuated through resistors; photovoltaic panels can be added; and the energy stored may be used to avoid buying from the grid during peak hours.

The Arrowhead elevator demonstration[7] aims at designing a smart energy system for elevators (and possibly other machines), leveraging energy recovered from the elevator, solar energy produced locally, and local energy storage, while being connected to the grid. The main difference with other multisource energy systems comes from the use of connected services in order to optimise the control of the system according to cost and CO_2 criteria. The management

---

[7]Involving Sodimas, Grenoble INP, Schneider Electric and the CEA. The reader is referred to [13] for an early overview of the demonstration.

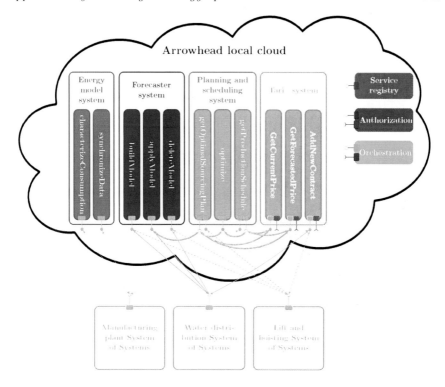

FIGURE 7.9
Service-Oriented Architecture of Schneider applications.

of energy sources and storage relies on forecasts of energy demand, of energy recovered from travels, of solar energy production, and of grid energy price. The forecasts feed an optimiser that provides a local energy hub controller with the best strategy to follow. This information is regularly updated (for example every 15 minutes).

The principles of the lift scenario are the following:

- The elevator and its energy system are equipped with sensors. The elevator sensing interface regularly provides information on elevator travel characteristics, energy production and consumption, solar energy production, battery status, etc. This information is sent to the cloud, gathering the data and hosting prediction and optimisation services, as described below

- Based on historical data, the DemandPrediction service is used to forecast the energy production and consumption of the elevator for the next period, typically 24 hours in time steps of 15 minutes

- The WeatherForecast service is used to forecast the irradiance and hence the production of the solar panels throughout the period under consideration

- The EnergyTariff service is used to retrieve the electricity tariff throughout the period.

- The StrategicOptimiser service is used to plan the usage and storage of energy throughout the period. It relies on a linear programming model presented in [14]. For each time step, the plan specifies how much energy is produced or consumed by each component. In particular, it indicates how much to buy from the grid, and the state of charge of the battery at the end of the time step.

- The LocalEnergyHubController allocates energy in real time, depending on the elevator travels, as much as possible in accordance with the strategic plan.

- When needed or at regular intervals of time, the StrategicOptimizer is called again to revise the plan, based on the information sent to the cloud through the elevator sensing interface.

At this time, the evaluation of benefits (savings incurred by using the overall solution and potential return on investment) is ongoing. First, there are benefits associated with the use of batteries, solar panels, and energy recovery. These benefits are related to energy costs and increased availability of the system in case of grid failures (a necessity in some buildings hosting people with reduced mobility). In addition, there are benefits associated with the use of prediction and optimisation. A first evaluation (with a simulated elevator and using historical data regarding travel) provided encouraging results, with electricity costs being about 35% of those resulting from a naive control system. Combining the two sources of savings, we estimate that up to 65% of the electrical costs could be saved, depending on the context (weather, occupation of the building). We still need (i) to refine these estimations and understand how they depend on the elevator usage, on the external conditions, and on the quality of the forecasts and (ii) to evaluate the return on investment for a number of alternative options.

The capacity of the elevator energy system to participate in a virtual market through FlexOffers is also under investigation. The objective is to describe the flexibility offered by the presence of the battery and assign an economic value to the activation of this flexibility.

7.4.2   Smart elevator monitoring and control

Buildings are one of the most important sectors with a significant potential for improving energy efficiency. Currently the residential sector alone accounts for 30% of the electrical energy consumed in OECD countries, corresponding to 21% of energy-related CO_2 emissions. The International Energy Agency estimates that energy-efficiency improvements could achieve a reduction of 47% in energy-related CO_2 emissions by 2030. At present, buildings contain an increasing number of installations, and among them are vertical transportation systems. Energy consumed by elevators represents about 5% of the total electricity consumption in buildings. Even if this amount may not seem significant, a 30% to 60% efficiency gain may lead to a relevant energy reduction, if we consider all elevators in operation all over Europe (4,800,000 elevators in UE-27). Aware of this scenario, elevator companies have oriented their R&D work towards development of intelligent, sustainable and safe transport systems that are energy efficient and have a better integration in buildings. To validate Arrowhead results, a full-scale pilot has been deployed in an energy aware building located in the IDEO technology park. The pilot consists of five buildings: ORONA Foundation building with four elevators, A3 Building with one elevator, ORONA IDEO Gallery with one elevator and ORONA ZERO with six elevators. Three elevators integrate Renewable Energy Sources (RES) (photovoltaic systems), four of them use an electric storage system and an energy management system, and all of them incorporate smart-meters, are enabled to return energy to the grid, and are managed by a central monitoring and supervision system.

7.4.2.1 Arrowhead approach

A major requirement concerning services in a full scale pilot experiment is to guarantee compliance with the IEC61850 standard, which is about communication and information technologies to ensure interoperability in future grids. This standard - originally aimed at substation automation, has been expanded to cover the monitoring and control of Distributed Energy Resources (DERs). A smart elevator can be considered a DER system, which in turn consists of smaller DERs such as the elevator itself, the integrated RES, the battery and the smart-meter. A smart elevator is usually a consumer of energy, but it can also act as an energy generator (when travelling down) and as power storage. Smart elevators can exchange power with the grid, acting as energy producers or consumers, depending on different variables such as the state of charge of the associated battery, or the current price of energy.

In order to guarantee compliance with both the Arrowhead Framework and IEC61850, an assistant tool has been used. That tool, developed by IK4-IKERLAN, facilitates the process of generating IEC61850 compliant Arrowhead Adapters - the middleware between legacy systems and Arrowhead Framework systems. This assistant tool facilitates the creation of Arrowhead-IEC61850 information models through a graphical user interface.

Arrowhead compliant services developed for the smart elevator system have been classified into three groups:

a) IEC61850 core services, which implement core IEC61850 concepts, such as authorisation and reporting

b) Basic IEC61850 compliant services, which mirror IEC61850 concepts such as logical devices, logical nodes, datasets, and report control blocks

c) Highlevel services, built on top of previously mentioned services, to implement advanced applications such as monitoring, data logging, energy efficiency management, maintenance, configuration, calibration, metering, etc.

For registration purpose the Arrowhead ServiceRegistry is used. Concerning authorisation, an Arrowhead compliant and propietary services is used.

Figure 7.10 presents the services implemented in full scale pilot, two IEC61850 core services, six basic services and two high level services.

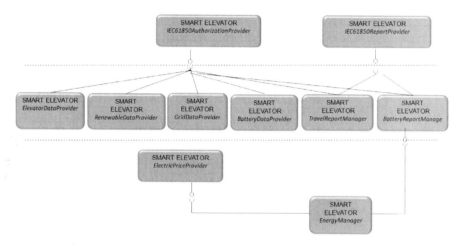

FIGURE 7.10
ORONA IDEO Smart-elevator pilot.

7.4.2.2   Benefits

The main benefits for users and vendors resulting from application of the Arrowhead approach to smart elevators are:

Benefits for users: Energy Storage Systems (ESS) integration on elevators will increase their efficiency between 15% and 25%%[1]. This improvement will assure a similar economic saving for the user. ESS will reduce peak power demand from the main grid, contributing to a 20–25% reduction of the power cost part of the electricity bill. RES connected behind the meter will reduce the dependency on the main grid and achieve savings of 50% on the electricity bill. Local and remote real-time monitoring will contribute to improving the elevator's assistance service quality and facilitating maintenance works. ESS,

---

[1]Information based on ORONA internal analysis.

RES, and monitoring/management system integration will increase the traditional elevator cost by 20%, but this extra cost could be easily recovered in 7 years.

Benefits for vendors: Validation of near Zero Energy Lift (nZEL) will contribute to identifying new technical improvements and to acquiring practical know-ledge about RES and ESS applications at elevator level. These goals will provide more flexible and better products - in addition to well-sized ones, to cope with energy requirements of the smart elevator. Higher levels of efficiency will be achieved, guaranteeing sustainability. At first, the market strategy will be oriented toward the top level of the elevator sector. It is expected that sales will increase at least 5% due to the integration of innovative systems. ESS and RES cost reductions and technical improvements will open the market in 5 years. To sum up, ORONA expect to increase their sales in the elevation sector by improving efficiency, reducing energy demand, and preparing its elevators to participate actively in a scenario where the smart grid and buildings are singular actors.

Highlight: The Arrowhead project has enabled the validation of the smart-elevator concept proposed by ORONA. This new elevator will attain near zero energy consumption, integrating ESS, RES, and bidirectional communication systems. Thanks to these new systems, smart elevators will offer energy related services to buildings, such as load shifting and peak demand reduction. Smart elevators will also contribute to safer and reliable integration of renewable energy sources targeting building sustainability and meeting the climate change and energy policy objectives for 2020.

7.4.2.3 Exploitation of results

Energy efficiency is a top priority in the international agenda towards a more sustainable energy future. Indeed, according to the International Energy Agency, it is considered to be the most cost-effective concrete action that governments can take in the short term to address climate change and to reduce energy consumption.

Several institutional initiatives have been taken to improve energy efficiency in buildings. For instance, in 2002 the European Parliament developed the 2002/91/CE directive in order to promote energy efficiency in buildings. In April 2009, the European Parliament Industry Committee developed a report to reform the 2002 directive. That report proposed that by 31 December 2018 at the latest, EU Member States must ensure that all newly constructed buildings be Zero Energy Buildings (ZEB). A ZEB is defined as "building where, as a result of the very high level of energy efficiency of the building, the overall annual primary energy consumption is equal to or less than the energy production from RES on-site". At present European directive 2010/31/UE (EPBD recast) requires that states shall ensure that by 31 December 2020 all new buildings are nearly zero-energy buildings, and after 31 December 2018, new buildings occupied and owned by public authorities will be nearly zero-energy.

In this context the development of efficient, low consumption, and sustainable smart elevators will set up a business opportunity that will contribute to achieving new energy targets for nearly zero-energy buildings. Existing commercial product analysis shows that competitors have made great efforts to improve energy performance in their products. However, at present there are no commercial products that integrate RES, ESS, and an intelligent energy management system. Consequently, a fast and successful application of Arrowhead practices will be crucial to increasing competitiveness in the elevator sector.

7.5   Context aware streets

7.5.1   Intelligent urban lighting

Intelligent urban lighting is an integral part of the municipal environment, serving communities and local businesses, promoting economic development, and enhancing safety, security, and the aesthetic appeal of surrounding property (Figure 7.11. Municipalities generally install street lighting for practical reasons, and sometimes simply for aesthetics. Thus, the main objective of this urban lighting system revolves around the many ways in which the practical needs of light can be addressed by a single system.

The objective of the developed intelligent urban lighting system has been the reduction of energy consumption. To achieve this, monitoring the energy consumption of lighting equipment and the implementation of modules – which can act on the street lights to impose more dynamic control – has been a key functionality of the Arrowhead intelligent urban lighting system, deployed in the city of Barcelona by Indra.

Within an urban environment, areas with a high flow of people and vehicles are more prone to accidents at night and these drastically increase if inadequate lighting is in place or lighting intensity is poorly distributed or controlled. These issues get more problematic in urban areas as the infrastructure gets older. Lighting management systems, lamps, and roads, for example, become outdated roughly at the same time and can jointly pose serious safety issues. As such, the main objective of this urban lighting system is to explore the benefits obtained by combining street and building environment information with mobility information to illuminate the streets according to their real-time needs. With this aim, Indra has designed a system to monitor and control each urban lighting element in real time based on its experience designing and developing monitoring and control applications.

The building environment is provided by a SmartBuilding service, also developed by Indra, which is integrated with the framework and makes energy consumption data available to other services, among which is the Intelligen-

FIGURE 7.11
Picture of the street where intelligent urban lightning has been deployed.

tUrbanLighting service. Thanks to the information published by this service, the UrbanLighting system can take into account the building usage patterns to determine the best street illumination.

The design of the smart lighting application is modular, consisting of five modules with clearly defined aims. The first module, the core, which integrates a rules engine, is responsible for actuating against the street lights based on input data received from other services and various sensors. The second one is the web component that makes it possible to customise the behaviour of the previous module and to monitor and actuate over the street lights directly.

In order to communicate with the Arrowhead Framework, the system relies on two other modules, one for obtaining data from a smart building system whose information will be used to determine the needed light intensity, and a second one for publishing power consumption information as an Arrowhead compliant service so it may be used by others. The last of the five modules is the on-street control unit which receives information from the core and uses wireless technology to change the intensity of the street lights. It is also responsible for sending information to the core about the state of the street lights and other sensors.

This last module also acts as a safety measure in case communication with the core is lost. In such cases, it would change the intensity of the street lights to ensure good visibility and thus the safety of the area.

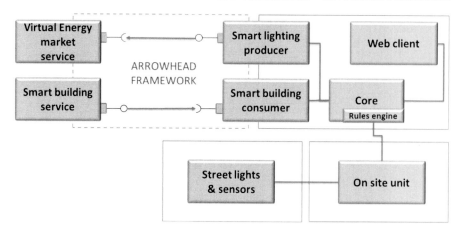

FIGURE 7.12
Intelligent urban lighting – Design Diagram.

This application has been designed to obtain context data through the Arrowhead Framework.

As stated before, one system module is responsible for this. It enables the system to integrate into the Arrowhead Framework, enhancing the capabilities of the system. The connector module itself is also designed to be customizable and its design has evolved during the pilot.

Firstly, the information from the smart building service was received from a Java Message Service (JMS) broker and the Arrowhead framework was used to obtain the connection parameters to the previously introduced messaging system.

In a second, more mature design, the connector retrieves information from the smart building using the REST capabilities provided by the framework, making the overall system more integrated and capable of taking advantage of other framework features, see Figure 7.12. Similarly, the producer is responsible for making information from the smart lighting available through the framework, and also makes use of the REST technology to publish information.

This urban lighting solution has been designed to consume street context data from other Arrowhead services in order to carry out smart control of the lighting intensity, see Figures 7.13 and 7.14. This lighting application has been connected to Smart Building Services to obtain data related to the buildings. Thus, the goals of increasing the perception of safety and security as well as improving pedestrian visibility are also incorporated in this urban lighting system.

FIGURE 7.13
Intelligent urban lighting application deployed by Indra in Barcelona.

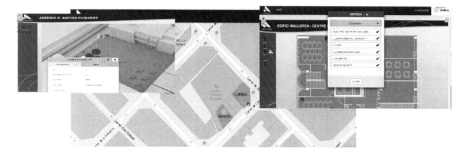

FIGURE 7.14
Intelligent urban lighting integrating context data through the Arrowhead Framework.

7.5.2 Intelligent car heating

Preheating of car engine in cold weather reduces fuel consumption and exhaust gas emissions. In Nordic countries, electrical heating poles are widely used for preheating the car engine and/or car interior at parking yards in winter time. Traditional heating poles have a manual on/off switch or 24-hour timer to activate the heating for 2 hours. The drawback of the traditional systems is that the heating time is not adjusted based on temperature; thus heating too long consumes unnecessary energy. For saving energy and providing users with better comfort, the heating pole should have remote control capability for user comfort and energy efficiency. This section describes an intelligent car heating pole that can be controlled and monitored using Arrowhead compliant tools.

The heating pole hardware and embedded control has been developed by the Arrowhead partner THT Control Oy (Finland). The heating pole (Figure 7.15 left side) has an aluminium base and contains two to four electric sockets depending on a model. The heating pole can be remotely controlled and monitored. The user has a mobile phone application for manually turning on and off the heating and creating automated heating schedules for different days. For the schedules, the user needs to inform only when the car should be ready and the heater calculates the energy efficient heating duration. The heating pole is connected to a server for storing information of all heating poles and their schedules.

The Arrowhead application service description and implementation were created for the car heating system by Tampere University of Technology. This interface is used for monitoring the car heater energy consumption and other parameters, and it can also be used for control purposes. It provides an Arrowhead service producer with a RESTful interface, which offers an access to car heater from the Arrowhead local cloud. The heater service producer uses ServiceRegistry and Authorisation core services. The ServiceRegistry is used to publish the endpoint of car heater thus making it discoverable for other systems in the Arrowhead local cloud. Furthermore, the Authorisation system is used to verify that the consumer is authorised to use the services of the car heater. The Orchestration system can be used for connecting other systems automatically to the car heater.

The pilot implementation is composed of the heating pole, street light, additional temperature and light sensors (Figure 7.15). Each system is described with Arrowhead service descriptions. The external temperature sensor is used for providing accurate outdoor temperature to the car heating system. The light sensor is used to provide brightness information to control the street lights and the lights in the car heater.

The implementation demonstrates the use of the Arrowhead Framework in smart city applications, allowing for a seamless integration of different company products (car heater, streets lights). The context awareness of the system allows the use of third party temperature and brightness information to achieve energy efficient heating and lighting. In the future, the knowledge of user behavior (e.g., scheduled heating times) could be used for adjusting lighting of the street lights.

7.6   Optimisation of municipal service systems

Municipal services (heat and water distribution, building services, etc.) require effective monitoring and control of heat energy consumption to prevent thermal energy loss in district heating networks, undetected accidents, hidden leakages, and inefficient utilisation in water distribution networks. A range of

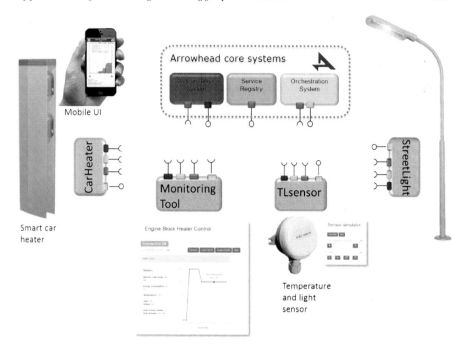

FIGURE 7.15

Smart car heater demonstration. Car heater is integrated into other systems using Arrowhead Framework.

information and communication technology systems, called municipal service systems (MSS) are deployed to handle this monitoring and control.

MSSs in Baltic countries use automated meter reading (AMR), smart meters, and other methods for collecting metering data from sensors installed in heat, electricity, and water distribution networks. However, there are several practical issues preventing introduction of smart meters: underestimated cost of technical solutions, frustration in selecting relevant solutions for local needs, lack of technical competence, and incompatibilities between protocols used in legacy systems and new equipment offered by suppliers.

Despite being owned by the same entity (the municipal council), each municipal service maintains its own network of meters and sensors, its own system for data collection and storage, separate customer service, inventory, bookkeeping, billing, etc. The majority of the systems are obsolete and incompatible. The aim of this research is to provide evidence of a practical implementation of a new generation of automated network monitoring systems for use by municipal services, which comply with Arrowhead framework approach:

- Development of SOA based application services as web services applying most suitable IEC standards.

- A modular technical solution for sensors and gateway nodes.

- Develop a concept of common core services for the cloud of MSSs.

7.6.1 MSS cloud description

This implementation aims at designing a System of Systems (SoS) that comprises enhanced MSSs, to create and drive application services produced by these systems (see Figure 7.16). The systems considered are

- Water supply system

- District heating system

- Building maintenance system

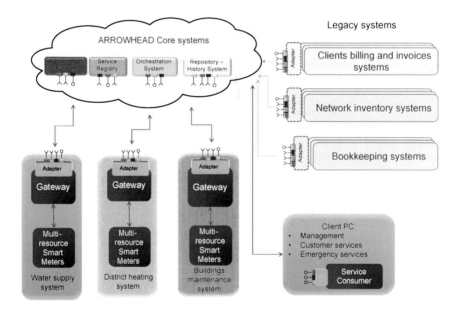

FIGURE 7.16
Municipal service system architecture

Furthermore, an SoS, which can be recognised as a cloud of MSSs, includes several legacy systems, such as network inventory systems, client billing and invoicing systems, and bookkeeping systems, which are operated by several of the service companies owned by the municipal council. Additionally, client

PCs, are recognised as a part of the SoS, since they behave as service consumers.

For the purpose multiresource meters developed and tested. These are suitable to apply to water distribution networks, district heating substations and temperature and humidity measurement in the offices and apartment buildings. Furthermore, a modular solution for gateways has been developed. The gateways also implement adapters and protocol translation functions for communication with other systems. Thanks to the enhancement of sensors and gateway devices, it became possible to design and implement a common architecture for MSSs partly realizing the Arrowhead Framework approach.

Heterogeneous MSSs apply different standards. For example, ISO 4064-1:2014 specifies the metrological and technical requirements for water meters for cold potable and hot water. These water meters incorporate devices which indicate the integrated volume. Many electricity meters are equipped with an IEC1107 compliant optical interface. This gives a convenient method for electricity companies to access information held in the meter using a hand terminal.

We considered also IEC 61850, a communication standard for electrical substation automation systems. The abstract data models defined in IEC 61850 can be mapped to a number of protocols. These protocols can run over TCP/IP networks or substation LANs using high-speed switched Ethernet to obtain the necessary response times below four milliseconds for protective relaying. However, IEC 61850 offers too complex models and features for simple scenarios maintained at the MSSs.

IEC 62056 is a set of standards for electricity metering data exchange by the International Electrotechnical Commission, IEC. We have proposed a mapping of the IEC 62056 standard functionalities to web services. The IEC 62056 standards are the International Standard versions of the Device Language Message Specification/Companion Specification for Energy Metering (DLMS/COSEM) specification. DLMS is the suite of standards developed and maintained by the DLMS User Association that has been adopted by the IEC TC13 WG14 into the IEC 62056 series of standards. COSEM includes a set of specifications that defines the Transport and Application Layers of the DLMS protocol. More information can be found in [15].

The IEC TC13 WG14 groups the DLMS specifications under the common heading: "Electricity metering data exchange–The DLMS/COSEM suite." The DLMS/COSEM protocol is not specific to electricity metering; it is also used for gas, water, and heat metering:

- IEC 62056-5-3:2013 DLMS/COSEM application layer

- IEC 62056-6-1:2013 Object Identification System (OBIS)

- IEC 62056-6-2:2013 COSEM interface classes

The COSEM server model is structured in three hierarchical levels: physical device, logical device and Accessible COSEM objects. Therefore, we have proposed a mapping of the IEC 62056 standard functionalities to a RESTful approach. As the paths resemble a fully qualified file name notation, they can be mapped to a URL, which makes the data model suitable for web services; a URL could look something like `http://hostname/device/node/class/attribute`.

The proposed solution for communication as a part of MSSs contains a multiinterface modular platform with two main node components. Metering nodes connect to a meter via switchable/selectable interfaces (current loop, IEC1107 optical interface, etc.). Metering nodes have rechargeable batteries for operation during a power outage and for long-term standalone operations. Inter-system communication is possible using selectable interface modules (e.g., IEEE 802.3, ISM radio interface, GSM/GPRS). Gateway nodes have selectable inter-system and back-end communication interfaces. A gateway node provides requests, readout preprocessing, secure data delivery, queuing, etc.

The registration of gateways can be performed automatically or manually, with the following parameters specified: device name, group, type, network settings, gateway dynamic loadable modules, VPN configuration, Wi-Fi settings, encryption settings, the working mode registration server/service (WMRS), and working mode data submission server/service parameters (WMDSS). The WMRS and WMDSS parameters allow the gateway control functions to be delegated to other servers for load balancing or to third party service operators by allowing them to perform all parameter and control functions as the functions transfer registration service (FTRS).

All gateways initiate an SSL VPN based on FTRS or WMRS VPN setting parameters to the data processing service that receives the gateway data through the encrypted VPN tunnel and stores them in a database. For the prototype application and further integration into any of the third party networks, a service agent at the data processing service or the gateway itself has to be integrated. In the prototype case, a web service demonstrating configuration and data reception/visualization is used. Querying has been implemented using web services.

The described communication part of MSSs has been deployed at the water supply network in the city of Ventspils (Water utility "Udeka").

7.6.2 Application services

Four Arrowhead Framework compliant application services have been created as web services:

1. Water consumption (Water_http)

2. Heat energy consumption (Heat_energy_http)

3. Temperature (Temperature_http)

4. Humidity (Humidity_http)

The first is produced by the water supply system, the second by the district heating system, and the last two by the building maintenance system. All services are hosted at the Smart Meter server centre `http://arrowhead.` `bitdev.lv/admin`. Therefore, the enhanced MSSs are able to consume published services inside of the local MSS cloud. However, potential service consumers outside of the MSS cloud will be able to consume application services through the usage of the Gatekeeper service provided by the core Arrowhead Framework.

7.6.3   Further development

The mandatory core systems such as Authorisation, and Orchestration have been used as provided by the open Arrowhead clouds provided by Arrowhead project partners BnearIT and Luleå Technical University. In addition, part of the functions of Orchestration system have been tested at Ventspils water distribution network, namely, gateway and sensor nodes registration and reconfiguration.

A data repository for each MSS will make use of the core Arrowhead Framework Historian system. It has been integrated to the MSS local cloud and piloted at Ventspils water distribution network and for building maintenance systems in Riga. Sensor data related to water flow, water pressure, temperature, heat consumption, humidity in the building etc., using encodings like JSON and XML, are stored, automatically decoded and processed.

The data are automatically recorded with their time stamp and metadata. Decoded sensor data can later be processed and filtered for analysis and visualisation purposes. The encoded data and its visualisation are available for client's PCs.

Such data needs a particular semantics and visualisation support: for example, management services benefit from aggregated data shown as tables and charts, which are refreshed by day, week of month. On the other hand emergency services need data to network monitoring system at least each 10 minutes (Ventspils water utility). The data repository–History system further provides services to client's PCs via a web interface.

A challenge still is services provisioning to the legacy systems. It is planned to develop an adapter and to test it on at least one of the systems. As a preliminary solution, a semi-automated translator was tested between the data repository and the billing system at Ventspils water utility, which converts data in a required format. The billing system operator calls the data repository and the service loads data to the billing system on request.

7.6.4   Advantages and savings

Since municipal services nowadays pursue automatisation of the processes but install and maintain very diverging systems, the issue of maintenance cost optimisation becomes very crucial, and particularly, when the maintenance cost causes growth of tariffs for the municipal services. Application of the Arrowhead Framework approach for optimisation of municipal service systems demonstrates promising opportunities for systems maintenance cost saving.

- Opportunity for sharing of application services among municipal service companies may reduce requirements for functionality of their systems, for example, if one of the systems provides a temperature service, it is not necessary to measure and collect temperature data by the other systems at the same territory.

- Since the information describing the different tasks and configurations is stored in a network connected storage area, i.e., in the core system accessible from all locations, where devices are to be installed and in a format that the devices are able to interpret, network nodes installation, configuration, and reconfiguration process becomes faster and less time consuming. Additionally, the cost of transport for technicians visiting the sites can be reduced.

- The data provided by the data repository–History system, filtered and processed in a required dimension, granularity and frequency, is an invaluable asset for Client PCs and other specialised systems, which interpret and use them. For example, legacy systems such as SCADA, EPANET, etc., can effectively utilise data for damage, leakages, burst of pipes, or theft cases identification and localisation.

- Acquisition of the Arrowhead Framework core services for the MSS cloud encourages a transition from multiple legacy systems, which belong to different utility companies, to the restricted number of shared systems that are able to collaborate with the Arrowhead Framework core and application service networks. For example, only one billing system instead of three or four is sufficient to serve the same client base.

At this time, the evaluation of benefits (savings incurred by using the overall solution and potential return on investment) is ongoing; however, the first evaluation of the saving by different municipal services looks encouraging.

7.7   Conclusion

In this chapter, we have introduced a series of applications of the Arrowhead Framework on energy management, provision, and optimisation. The applications include solar powered lifts, car heating, intelligent lighting of streets, and a virtual market of energy. First of all, the examples show that it is possible to expose such applications as standardised services by extending them with simple add-ons and by using moderate effort. Secondly, they indicate how the

Arrowhead Framework may be used in future ICT infrastructures of Smart City/Smart Grid systems.

Bibliography

[1] L. L. Ferreira, L. Šikšnys, P. Pedersen, P. Stluka, C. Chrysoulas, T. L. Guilly, M. Albano, A. Skou, C. Teixeira, and T. B. Pedersen, "Arrowhead compliant virtual market of energy," in *Emerging Technology and Factory Automation (ETFA), 2014 IEEE*. IEEE, 2014, pp. 1–8.

[2] T. L. Guilly, L. Siksnys, M. Albano, P. Pedersen, P. Stluka, L. L. Ferreira, A. Skou, T. Pedersen, and P. Olsen, "An energy flexibility framework on the internet of things," in *The Success of European Projects using New Information and Communication Technologies*, S. Hamrioui, Ed. Scitepress, 2016.

[3] L. Šikšnys, E. Valsomatzis, K. Hose, and T. B. Pedersen, "Aggregating and disaggregating flexibility objects," *IEEE Transactions on Knowledge and Data Engineering*, vol. 27, no. 11, pp. 2893–2906, Nov 2015.

[4] P. Saint-Andre, "Extensible messaging and presence protocol (xmpp): Core," Internet Requests for Comments, RFC 6120, March 2011. [Online]. Available: http://www.rfc-editor.org/rfc/rfc6120.txt

[5] M. Albano, L. L. Ferreira, L. M. Pinho, and A. R. Alkhawaja, "Message-oriented middleware for smart grids," *Computer Standards & Interfaces*, vol. 38, pp. 133–143, 2015.

[6] "Openadr alliance," http://www.openadr.org/, accessed: 2015-11-28.

[7] X. I. W. Group, "P21451-1-4 standard for a smart transducer interface for sensors, actuators, and devices based on the extensible messaging and presence protocol (XMPP) for networked device communication," IEEE Standard Association, IEEE, 2008.

[8] R. T. Fielding and R. N. Taylor, "Principled design of the modern web architecture," *ACM Transactions on Internet Technology (TOIT)*, vol. 2, pp. 115–150, 2002.

[9] P. Saint-Andre and T. Alkemade, "Use of transport layer security (tls) in the extensible messaging and presence protocol (xmpp)," Internet Requests for Comments, RFC 7590, June 2015. [Online]. Available: http://www.rfc-editor.org/rfc/rfc7590.txt

[10] M. Albano, R. Garibay-Martínez, and L. L. Ferreira, "Architecture to support quality of service in arrowhead systems," in *INForum-Simpósio de Informática (INFORUM 2015)*, Covilhã, Portugal, Sep. 2015.

[11] K. M. Nielsen, P. Andersen, and T. S. Pedersen, "Aggregated control of domestic heat pumps," in *Proc. IEEE Conference on Control Applications (CCA) 2016.* IEEE, 2013, pp. 302–307.

[12] M. Louvel and F. Pacull, "Linc: A compact yet powerful coordination environment," in *Coordination Models and Languages: 16th IFIP WG 6.1 International Conference.* Springer, 2014, pp. 83–98. [Online]. Available: http://dx.doi.org/10.1007/978-3-662-43376-8_6

[13] V. Boutin, C. Desdouits, M. Louvel, F. Pacull, M. I. Vergara-Gallego, O. Yaakoubi, C. Chomel, Q. Crignon, C. Duhoux, D. Genon-Catalot, L. Lefevre, T. H. Pham, and V. T. Pham, "Energy optimisation using analytics and coordination, the example of lifts," in *Emerging Technology and Factory Automation (ETFA), 2014 IEEE.* IEEE, 2014, pp. 1–8.

[14] C. Desdouits, M. Alamir, V. Boutin, and C. L. Pape, "Multi-source elevator energy optimization and control," in *European Control Conference (ECC), EUCA 2015.* EUCA, 2015.

[15] M. Albano, L. L. Ferreira, and L. M. Pinho, "Convergence of smart grid ict architectures for the last mile," *IEEE Transactions on Industrial Informatics (TII)*, vol. 11, pp. 187–197, 2015.

8

Application system design - Maintenance

Erkki Jantunen

VTT

Mika Karaila

Valmet

David Hästbacka

Tampere University of Technology

Antti Koistinen

Oulu University

Laurentiu Barna

Wapice

Esko Juuso

Oulu University

Pablo Puñal Pereira

Luleå University of Technology

Stéphane Besseau

Airbus

Julien Hoepffner

Airbus

CONTENTS

In this chapter the use of Arrowhead Framework is tested in two application environments. The first part of the chapter "Widening the scope: Managing maintenance data in large multifunctional plant environments" reports the challenges in the mining industry and how the challenges of this type of industry can be addressed with the Arrowhead Framework. The second part of the chapter "Aircraft Maintenance System" is from commercial aviation focusing on various examples related to maintenance.

8.1 Widening the scope: Managing maintenance data in large multifunctional plant environments

In this section the use and role of the Arrowhead Framework in large multifunctional industrial automation environments is discussed. The basic scenario is related to enabling industrial IoT and connectivity of devices in a process industry setting. The basic motivation for the Arrowhead Framework - to facilitate the interoperability of IoT devices - is discussed in the light of practical pilot examples. The pilots are from the mining industry focusing on various examples related to maintenance. These examples were chosen due to the frequent wear in mining production equipment and the difficulties associated with traditional condition monitoring for production equipment scattered over long distances. Consequently, the new IoT solutions for gathering and analysing data in the mining industry have high potential for lowering maintenance costs and increasing the availability of production equipment.

The chapter describes how the Arrowhead Framework can be seen as an open-source proof-of-concept prototype of a self-configuring IoT sensor system framework suitable for industrial process control environments and also for supporting maintenance processes.

8.1.1 Introduction (scenario)

The investigation started with a number of possible targets that should be monitored at Outokumpu Chrome Oy Kemi Mine: hoisting machinery rope damage detection, conveyor pile level measurement, and wear plate condition measurement (metal plate thickness). In the first phase all these cases were studied and feasibility studies with initial hardware were carried out. After some brainstorming workshops at Kemi Mine and studies at the actual process environment, a number of feasible solutions were selected.

The main focus was on providing a system with an acceleration sensor with Bluetooth LE connectivity to monitor the wear of wear plates in the conveyor. For comparison, a high-end validation system was also installed.

Online vibration analysis methods were tested on the rod mill at Kemi Mine's concentrating plant. The rod mill is a primary refiner in the grinding circuit, taking the maximum power of 1800 kW and so being the greatest energy consumer in mineral processing. Down time and non-optimal operation creates huge expenses in terms of lost production and energy with the mills of this magnitude. Even small improvements in operation and reductions in down time will bring significant savings [1].

In the tested measurement system, vibrations caused by the monitored event define the requirements for the logging arrangement. Vibration logging requires the measurement setup to be capable of recording data at sampling frequencies at least twice as high as the highest monitored frequency. Vibrations were recorded from trunnion bearing supports with four SKF CMSS 2200 accelerometers and data logging was performed with National Instruments cRIO-9024 controller with a cRIO-9114 chassis that includes a Xilinx Virtex-5 reconfigurable FPGA core. Vibration sensors were connected to an NI 9234 analog input module with a built-in anti-aliasing filter.

For storing measured data, remote cloud data storage was set up according to the Mimosa Open O&M information model [2]. For easy connectivity from devices and gateways a REST interface was developed, also usable by value-adding service wanting to further analyse the data as well as end-user applications. Additional control system connectivity was under discussion using OPC UA [3] and Wapice's OPC UA REST API was used as one solution.

The pilot focused on the following main points

- IoT and interoperability of IoT devices

- Arrowhead Framework developed to facilitate interoperability of IoT devices

- Automatic device detection

- Custom sensors and their use in a factory environment:

  - Hoisting machinery rope damage detection (camera & image analysis)
  - Test plate thickness measurement
  - Vibration analysis logging
  - Automatic logging and (possible) readjustment of machine settings

- Objective: To develop an open-source proof-of-concept prototype of a self-configuring IoT sensor system that utilizes the Arrowhead Framework

8.1.2   Motivation for monitoring

The ability to include relatively efficient processing power in small sensors and embedded devices in the field opens many possibilities for predictive maintenance and operational monitoring [4]. IoT solutions bring the information from numerous sensors together and enable interactions between distant devices and systems. The possibility of using this new refined information can shorten the downtime in maintenance breaks by enabling prognostics through the ability to calculate the remaining useful life (RUL) of monitored components. It can also be used for signaling failure alarms to maintenance organisations and spare part providers for fast response. Relevant metadata can also be connected with the measurement information so that the maintenance operators have all the needed information even before consulting the crew at the plant. These kinds of means can enhance operator and maintenance personnel performance with facilitated access to relevant information [5].

In industry the production machinery suffers from wear and consequently maintenance is needed in order to keep the production machinery running at high efficiency. In principle the wear of components of machinery can be monitored based on time, load, or wear, but in practice various factor may have a significant influence on the analysis outcome.

8.1.2.1 Time-based monitoring of wear

Time-based monitoring of wear which would support time-based maintenance is not very efficient because wear does not take place as a function of time, i.e., of the four main models of wear (abrasive, adhesive, fatigue, chemical) only the chemical one can be considered to be a function of time. The others types of wear only take place when the machines are running. In spite of the fact that it is easy to understand that if a component of the machinery is on a table beside the machine it does not wear in a similar manner as inside the running machine, time-based maintenance is widely used. The reason for the wide use is naturally the ease of use, i.e., only a calendar is needed.

8.1.2.2 Load-based monitoring of wear

The loading of the components of machinery very much influences wear. In fact, doubling the load might increase the wear by an order of magnitude, i.e., making it ten times more severe. The loading of the production machinery can be either stable or varying in nature. If the load stays constant, the wear of the components can be monitored through monitoring the running hours, i.e., the time the machine has been used. When the loading of the machines varies the situation becomes more complex and both the loading and the time at a specific level of loading have to be recorded.

The higher loading of the machinery can, e.g., cause stress spikes which can start to form micro fractures in material after some time. Measuring and processing the raw data from these spikes into information can aid in estimating the developing condition [6]. The information extracted from the raw data can be treated like a process measurement such as a temperature reading. Extracted values do not include information about the severity, and, if the user does not know the meaning of these values, the measured information is of no value. This knowledge requirement can be solved by scaling the values so that the numbers are easily linked to the actual operating state. Nonlinear scaling is an effective tool for scaling these values to simple linguistic levels [very low, low, normal, high, very high] correlating with scaled value range of [-2, 2]. Change in condition increases the vibration levels so that the scaling function does not correlate with the changed situation. The scaling function can be updated recursively according to the changed condition in order to keep it up to date [7]. The linguistic levels can be used together with the traffic light presentation indicating severity with color coding. Simple severity levels with understandable descriptions and color coding ensure that these scaled values can be inspected and understood by anyone. The value scaling can be done at the local calculation node or at the connective node collecting several maintenance measurements.

8.1.2.3   Monitoring wear

Naturally, the most reliable way to define whether the machine needs maintenance is to rely on monitoring of wear as such. Unfortunately, the monitoring of wear can be technically very challenging and thus expensive. The reason for this is that the parts that wear are often moving and hidden by other parts or protective covers. Consequently, indirect methods such as vibration measurements are used to define the condition of the components of production machinery.

All of the machines with rotational or sequential operation generate vibrations according to their current state. These vibrations can be recorded and the data processed according to desired features about the machine condition or operation. Vibrations linked to certain aspect of machine operation are buried in the vibration signal with numerous other frequencies. Extracting the desired aspect of machine operation requires case-specific filtering and signal processing. Local real-time calculations can be used to extract information from the monitored vibrations without the need for demanding the transfer of the huge amount of raw vibration data to be processed elsewhere. Localised calculations can bring the extracted information to instantaneous use in automatic control or decision making. The long-term monitoring and tracking of cumulative stress gives valuable information for remaining useful life, RUL, calculations and prognostics, and short-term stress indices [8] indicate momentary high-stress periods, providing information about the causes of the wear and they can be used for control.

8.1.2.4   Condition-based maintenance

Knowing the condition of the components enables the use of condition-based maintenance (CBM) strategy, which has been widely accepted as the most efficient and economical strategy for carrying out maintenance. It is not the purpose of this chapter to go into depth about the benefits of various maintenance strategies. However, one extremely important aspect needs to be brought up, i.e., the adverse influence of carrying out maintenance actions in vain. Maintenance cannot usually be carried out when the machine is running thus unnecessary maintenance causes loss of production. Another even much more expensive result of unnecessary maintenance is the need to repair the consequences of previous maintenance actions in case something has gone wrong. From what has been explained here of carrying out maintenance in an efficient and economical way follows a great need for reliable measurements.

8.1.2.5 Maintenance in mining industry

In the mining industry the wear of production equipment is usually very high. This is due to the nature of processes since they often include such subprocesses as crushing and transport of rock and chemical treatment, both of which cause high wear. Consequently, optimally organised maintenance can have a large influence on the availability and efficiency of the production equipment. As explained in the previous paragraph, it is commonly accepted that CBM is the best strategy in this kind of environment. However, the big challenge in the mining industry is that the production equipment can be scattered over a large area, making it difficult and expensive to use wired sensors. Also in mines the sensors are exposed to accidents, i.e., something can hit them, thus breaking them, or something can cut the wires. Thus the wireless Bluetooth Low Energy (BLE) System on Chip (SOC) and the new Micro Electro Mechanical System (MEMS) sensors that are low energy and relatively cheap can give a real boost to the use of condition monitoring in mining industry. The above explanations are the main reasons for choosing the mining industry as the pilot and testing the new technologies at Kemi Mine.

In the conveyor pilot the aim is to follow the development of wear of a wear plate. The wear of the wear plate reduces the thickness of the plate which in turn lowers the stiffness of the plate to a greater extent than the weight loss which in turn reduces the natural frequency of the plate which can be monitored through vibration measurements. The purpose of this kind of monitoring is to be able to predict when the wear plate has to be changed and make this information available through the Internet to the subcontractors that then make offers of new wear plates. The pilot case can be seen as a rather typical opportunity to follow CBM strategy. The components that suffer from wear are monitored on-line and the development can be followed by a number of parties such as the plant maintenance personnel and a number of subcontractors who can provide spare parts or maintenance services. A more detailed explanation of the techniques that have been used to pass the data will be presented in this chapter. However, it should be noted that the solution is generic so that it could be used for any kind of condition monitoring measurement.

8.1.3 Data structure

In order to be able to carry out maintenance using CBM strategy as described in the previous paragraph, it is necessary to use a Computerised Maintenance Management System (CMMS [9]) since otherwise it would not be possible to keep track of planned and realised actions and the results of condition monitoring activities. There are a number of commercial CMMS packages available but most of them are oriented to the managing of maintenance work activities with limited support for condition monitoring and diagnosis activities that support CBM strategy. However, there exists an open solution for a plant description model and condition monitoring related data provided by MIMOSA ([10]).

8.1.3.1   MIMOSA

MIMOSA defines itself as a not-for-profit trade association dedicated to developing and encouraging the adoption of open information standards for operations and maintenance in manufacturing, fleet, and facility environments. MIMOSA claims that its open standards enable collaborative asset life-cycle management in both commercial and military applications. It is not the purpose of this chapter to go to detail about what MIMOSA can offer, but further information can be obtained from the web pages of the MIMOSA association ([10]).

8.1.3.2   Conveyor description

The relevant parts of the Kemi Mine conveyor have been defined in an SQL Server database using the MIMOSA Common Relational Information Schema (CRIS). It should be noted that this information schema would normally be used for the description of the whole plant but in this demonstration only the relevant parts of the conveyor and the components of the measuring system have been defined. In addition, all the measured condition monitoring and diagnosis data is saved in MIMOSA from which the developed service shows any chosen analysis values for any chosen assets (parts of production machinery).

8.1.4   Case-environment

To improve maintenance processes, new systems need to be developed that allow the transition to predictive maintenance based on condition monitoring data. Such systems consist of the following main components:

- Sensors

- Gateway servers

- Mediators

- Maintenance data stores

- Data-analysis services

- User interfaces

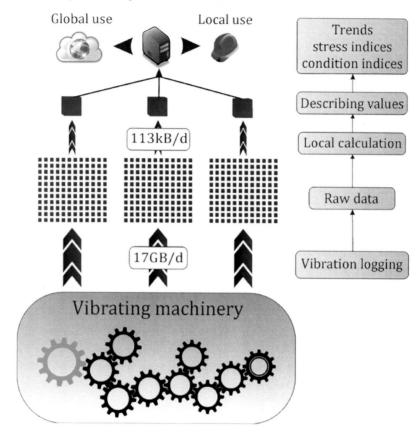

FIGURE 8.1
Local calculation for condition-based maintenance.

The recorded raw vibration data is needed for analysis purposes but a IoT solutions have no use for such dense data without connection to real-world events. Data collection should be planned according to possible future uses and balance between saving all of the data and discarding the data. Localised signal processing (see Figure 8.1) reduces the need for high-volume data transfer so that there can be a large number of nodes using the same transfer routes without exceeding the transfer capacity and the operation remains robust. Some applications may require the saving of the raw data from certain specific or exceptional events. In those cases there can be certain triggering for saving the raw data for some duration or until the end trigger occurs. Raw data can help in further analysis method development and historical data can act as a reference for future investigations.

8.1.5 Architecture

The Arrowhead Framework provides a ServiceRegistry (based on DNS-SD) that enables systems to find local services that they can consume and collaborate with. The service registry together with the other Arrowhead mechanisms of orchestration and security provide means for developing dynamic SOA applications that have a high potential for need-based scaling and configuration.

Systems providing services register an access point (e.g., REST API endpoint) automatically into the Arrowhead service registry. To keep the systems consistent it should also unregister when a service is shutting down. In the case of a crash or other unavailability it would be beneficial to use the time-to-live (TTL) parameter of the ServiceRegistry. This enables the removal of services not sending keep-alive messages during TTL. This will ensure that only services actually available are listed in the service registry.

Currently, for the applications presented in this chapter, the REST-based service registry interface is used for publishing and unpublishing services. This facilitates registry management and development because currently native DNS-SD tool support is limited in many modern runtime environments such as those based on C# and Node.js [11].

When a service instance goes online, it can first query for the REST service endpoint but currently this is provided as a parameter due to limited DNS-SD support in the platforms currently being used. The service registration is an HTTP POST request that includes the service name, type, domain, hostname, port as well as properties for version and path. For unregistering an HTTP POST to the unpublish path is performed with the name of the service as input.

To achieve dynamic behavior of service instances appearing and disappearing, a monitoring service type has been implemented that automatically removes monitored services if they disappear. It utilises the service discovery feature of the REST service registry to search for instances of specified service types. If a service instance stops responding for a given period of time, it is simply removed with the expectation that if it reappears the service will register itself. This means that when this monitoring service is also discoverable, services can use it to register themselves (or their type of services) to be monitored dynamically.

As gateways are connected and powered up, they will appear in the Arrowhead service registry. In the same way each Bluetooth sensor will connect to a gateway. The system scalability will be hierarchical as Arrowhead registry keeps gateway endpoints available and each gateway has own list of connected Bluetooth sensors. These sensors send data only when they have impact that will wake them up, otherwise sensors are in partial sleep mode and save battery. The gateway keeps all sensors and the last updated values in the buffer (cache). It can POST the latest data to the MIMOSA database (endpoint is provided by the Arrowhead Framework Orchestration system). The gateway provides a local user interface with the same endpoints as it provides data out.

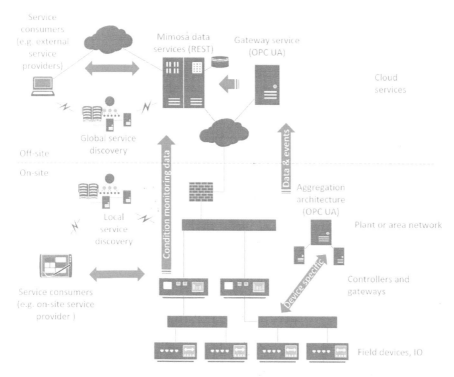

FIGURE 8.2
Overview of the general architecture.

8.1.5.1   OPC UA - REST API aggregation

OPC UA server interface

The OPC-UA aggregate server interface provides access to field data using the platform independent standardised OPC-UA protocol. Clients connecting to the server can browse the address space and subscribe to data nodes from any OPC-UA server. The main benefit of the OPC-UA based interface is the standardisation of the communication and the possibility to discover the address space and use subscriptions. Compared to other protocols often used in the field, OPC-UA devices additionally offer enterprise-level security and reliability on the application level. The use of aggregating OPC-UA servers in a hierarchy discussed in [12].

At cloud level, the downside of using OPC UA from other cloud systems is the requirement for a full OPC UA communication protocol stack. Developing an OPC UA based applications using only a protocol stack is quite a laborious task. Alternatively, commercially available software development kits and toolkits provide a more convenient starting point for a developer, but introduce additional costs. Such toolkits are available for various programming languages including Java and C#, common in cloud environments.

RESTful interface

The Representational State Transfer (REST) [13] is a style of software architecture for distributed systems. REST has emerged and established itself as a predominant Web service design model with key goals such as

- Scalability of component interactions

- Generality of interfaces

- Independent deployment of components

- Intermediary components to reduce latency, enforce security, and encapsulate legacy systems

An important concept in REST is the existence of resources (sources of specific information), each referenced through a global identifier (e.g., a URI in HTTP). In order to manipulate these resources, components of the network communicate via a standardised interface (e.g., HTTP) and exchange representations of these resources (the actual documents conveying the information). As a result, an application can interact with a resource by knowing two things: the identifier of the resource and the action required — it does not need to know whether there are caches, proxies, gateways, firewalls, tunnels, or anything else between it and the system actually holding the information. The application does, however, need to understand the format of the information (representation) returned, which is typically an HTML, XML, or JSON document of some kind, although it may be an image, plain text, or any other content.

In the aggregate server, cf. Figure 8.3, the REST interface provides access to the same data as the OPC-UA server interface but using a different protocol. Individual data nodes in the server are referenced based on a unique URI. The requested data is returned in a JSON or XML document. When the REST interface is used, the aggregate server operates as a bridge between these two technologies. The main benefit of using a REST-based interface is that there is no need for OPC UA SDK or Toolkit in order to access the data. Also, REST interfaces are very widely used Web technology, which is also convenient for the application developer. Therefore, a simple data access from a cloud application does not require significant development and the necessary tools and technologies are well supported in common programming languages such as C# and Java. One of the drawbacks of using a REST based interface is that the structuring of the resource URIs for the OPC UA data is not standardised. Additionally, some may consider the lack of a publish-subscribe type of communication scheme as a negative aspect of REST interfaces since REST clients needs to poll the data from the server. However the new HTTP-5 standard will provide the publish-subscribe capability.

Data model

A possible data model meant to simplify the use of different server interfaces and generically and logically reflect various organisational hierarchies is presented in Figure 8.3. At the root level of this system data model visible through the REST API are enterprises (e.g., companies or customers). Each enterprise can have multiple sites or locations of interest (e.g., a factory, a ship, or a fleet). Each site can define multiple assets or resources that need to be monitored (e.g., a device, an engine, or a truck). An asset can include multiple data nodes, describing what is measured from the asset it belongs to (e.g., temperature, engine RPM). An asset can also contain other assets further extending the model by creating a tree structure of assets. At the bottom of this data model, under each data node there can be multiple process data values.

All requests to the API are made through HTTPS. With HTTPS, the HTTP protocol is protected from wiretapping and man-in-the-middle attacks, therefore data being transferred is secure. Authentication to the API will be via HTTP Basic with username and password pair and should be sent along all requests that require it. Different system user profiles can be used to define the system level access to the data requested through the REST API.

The aggregate server could also provide other interface in addition to the OPC UA server interface and REST interface. These interfaces also do not exclude each other, and a different interface may be used in different situations based on the requirements of the application that is consuming the data. Other interesting interfaces for the aggregate server could be, for instance, message queue based interfaces.

8.1.6 Plant monitoring system components

8.1.6.1 Sensor and gateway devices

A Bluetooth low energy sensor is targeted to be low cost and energy efficient in such a way that it can be shipped with every wear part. It will work until the wear part has to be replaced. As the wear part is returned, the battery can be replaced and shipped again.

The sensor contains system on chip (SOC) that contains ARM Cortex-M0 CPU and BLE 4.0 Smart radio. The actual sensing component is an ST LIS3DH MEMS chip. These together are very low level energy consuming and energy consumption is optimised. The sensor will be active only when it gets impact that will wake it up. This threshold level can be configured locally or

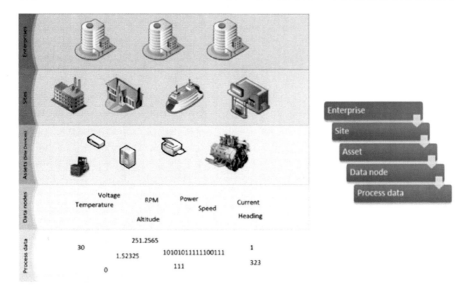

FIGURE 8.3
Aggregated data model.

from the cloud. The actual measurement contains raw values. The sensor will broadcast values and the gateway will store them and run FFT analysis.

8.1.6.2 Gateway server API

The first devices used were Raspberry B and BeagleBone Black; later versions of the gateway were tested with Intel Edison. All these are cheap and able to run Linux and small application (not enterprise size, main limitation is amount of memory). Each of these is used in many open source IoT examples and the same idea we had in using the commercial ready hardware as a gateway. Intel Edison with WiFi and Bluetooth connectivity was perfect a match for our requirements. It has a pre-built image with all software needed to run the applications.

The latest prototype fulfills basic measurement data storing locally and posting it to the MIMOSA database. It also has features to test the database connection and fetch stored data to the local user interface. The gateway application is based on Node.js and express web server. The management and orchestration were built with Node-RED [14]. It is a visual programming environment and contains a lot of ready nodes to implement application flows. The application connects to the gateway (GTW) and uses REST API to read sensors from the GTW. Each sensor can be configured from this user interface. As the sensor sends new measurement data, the application will get the data and then send it to the MIMOSA service and database. These communication messages are all based on REST APIs. The gateway provides all messages with Swagger documented and these can be browsed and tested through a web browser.

The last prototype will contain Arrowhead service registry binding. See Figure 8.4 for a small example of how the Arrowhead REST API can be used to list services and the service types and register a new service endpoint into the registry.

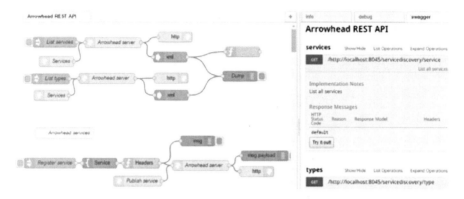

FIGURE 8.4

Node-RED applications contain REST API endpoints. At the same time the panel on the right shows and documents used operations. Users can test each operation immediately with the "Try it out!" button that Swagger provides if the API is documented in Swagger.

8.1.6.3 Mediators

The mediator pattern is used to work as a bridge between different protocols. The GTW runs Bluetooth LE protocol to get information from the sensors and then HTTPS REST API to send measurement data to the MIMOSA database. This separates low-energy sensors from the continuous connection to on-demand short time connections. As the GTW can hold (cache) data, it again reduces energy usage of the sensor.

The mediator pattern can be used with other protocols like CoAP to connect devices to the cloud. Even old legacy systems can be connected in this way. The vendor has to provide the node that can connect to the old system and give data in payload so that it can be formatted to the other protocol message format. In Node-RED this is easy to implement and to connect different systems with different protocols.

8.1.7   Results and further development

There are currently still some technical issues. The automatic configuration capabilities need to be enhanced. There has to be automatic service discovery via REST API. This is currently in beta-phase with the first integrated prototype now being implemented. Another foreseen modification need is the automatic white-listing process for sensor devices. This will make the sensor discovery faster. Related to this there is a need to expand to a more complete version and stress test the system with a larger group of active sensors.

In the last phase full integration with the Arrowhead Framework should take place. The authentication and orchestration services have not been used in the earlier prototypes.

The actual business potential will be seen in the future: what does the service framework bring? Does it facilitate engineering? Does it promote open systems enabling connectivity for new applications and services? Currently it can be used to connect services within and, to some extent, between clouds.

8.1.8   Discussion

When the Arrowhead project was started, there were a lot of ideas (too many actually) that have now been finalized into a vision of the final solution. There is a need for a dynamic system capable of devices and systems entering and exiting the service environment, i.e., those producing, storing and refining maintenance related data. For the end user there should be a user interface that can be used to configure sensors at the customer site.

User interfaces need to be browser based. In this way users can securely access any sensor at any site (globally, any cloud). Users can also browse from the same user interface available gateways and sensors. There is a default database at the gateway that is automatically selected from the available ones (discovered from the service registry). So the gateway is autonomous and it has registered itself into the service registry. This requires some metadata, because there can be multiple data storage services registered in the service registry. For example, a data service instance can be customer specific and can contain default parameters for each sensor. As the sensor is replaced, the gateway can reconfigure a sensor with the default parameters. This way systems run automatically with self-configuring sensors and gateways.

8.1.8.1   Application level issues

As already noted, the whole system scalability is not yet fully known since there are not enough sensors to run 50–100 sensors with one gateway. The gateway testing is not yet final as there is a need to stress test the system with several hundreds of gateways. Future work for expandability will cover integration with existing legacy systems. There are several standard protocols that should be integrated: OPC-UA, MODBUS-TCP, MQTT, etc. For the advanced, value creating services there should be analyzer services for end customer/service providers. These will need a notification service that can alert/indicate with loose interface like email or REST API to CMMS/ERP interfaces, e.g., create maintenance work orders. Alternatives to DNS-SD should be considered. For example multi-cast DNS-SD distributes information in the local network and contains local cache, better performance, security to be considered, and is more robust. An especially good generalisable and reusable part is the mediator pattern (no compile needed, Javascript).

8.1.9 Conclusion

In this section the features that have been developed and tested in the mining industry pilot have been presented. As such, mines are very demanding environments with significant potential for improvement of availability and efficiency. This environment has provided a good platform for widening the use of the Arrowhead Framework, linking it to a number of technologies and systems. All the tests between separate systems have been successful in the selected scale. The testing that has been carried out has taken place on such a small scale that it cannot be used as proof of a working solution that could be taken into wide use. Because the small-scale case testing, in some cases with very expensive equipment, has been successful, it is clear that the work should continue on a wider scale. As the Arrowhead Framework provides local - enterprise - cloud service registry, it can be used as a "yellow book" of IoT services. Mediators are needed to build protocol gateways that can provide interoperability between different systems and technologies.

As large and different systems can be very challenging to maintain, the system documentation should be based on online up-to-date descriptions. In our case we used Swagger v2.0 to show available endpoints and parameters of the provided REST API interfaces.

The Industrial Internet will actively grow in the next years, and the principles and solutions provided and tested in the Arrowhead Framework will be needed and used.

The Arrowhead Framework can be utilised to provide measurement data for third-party maintenance service providers and to make the information about comparable components easily accessible in a standardised manner.

8.2 Aircraft maintenance system

8.2.1 Introduction

Aviation is at the moment a fast growing market foreseen to double in size in the next decade, and it has lately seen a lot of changes with the arrival of new challengers with new business models leading to a great turmoil among legacy airlines and a higher level of competition. In this context, keeping aircraft flying is key for airlines as every unavailability can mean losing a lot of money, and aircraft maintenance is a noticeable contributor to aircraft downtime. Therefore, having better control of aircraft needs for maintenance and being able to react efficiently following a technical event is essential for airlines. That is the reason why Airbus, as a leading aircraft manufacturer, is continuously working at improving its products and at providing innovative and disruptive added value solutions for aircraft maintenance to support airlines in their day to day operations. The IoT concepts and their standardisation offers new perspectives in the way to approach system architecture and design that can help solve common issues in complex systems such as the aircraft maintenance system, and the Arrowhead framework offers to make it simple and effective.

8.2.2   Aircraft maintenance

8.2.2.1   Aircraft maintenance in a nutshell

Aircraft maintenance is the process of ensuring that an aircraft will be available and safe to operate its flights. In general, aircraft maintenance comprises two types of activities: scheduled maintenance activities and unscheduled maintenance activities.

Scheduled maintenance activities regroup a set of tasks (inspections, part replacements, tests, etc.) defined by aircraft manufacturers and validated by the aviation authorities that shall be performed by the airline to ensure aircraft continuous airworthiness. These tasks shall be planned by the airlines to ensure that they are performed on each aircraft in a given amount of time (calendar time, number of flight hours or flight cycles – a flight cycle being a take-off and a landing).

One example of scheduled maintenance activity can be the "daily checks" which consist of some light operations such as fluid levels control and visual inspections and can be performed when the aircraft is at the line station. There are also checks like "C checks" which consist of performing some thorough inspections and running some operational and functional tests on some components. Checks can require the aircraft to be immobilized for a few days or weeks in a hangar, which represents a prominent amount of time of unavailability of the aircraft for flight operations.

Unscheduled maintenance activities are maintenance activities that shall be performed specifically following the occurrence of an unexpected event in order to return the aircraft to safe operation. An unexpected event can be anything from a sensor or computer that failed, to a vehicle bumping the aircraft or the aircraft being struck by thunder, for instance. According to the impact of the defect on the aircraft, the aircraft operations, and the time available for maintenance, an unscheduled maintenance activity can consist of simply deferring the repair by setting the failed component into an inoperative state or making an inspection, or may require repairing or replacing a component or a structural part of the aircraft, which can have a huge impact on aircraft operations with flight delays or cancellations and everything that can result from such situation (passenger dissatisfaction, additional costs, etc.).

8.2.2.2 Challenges

Aircraft maintenance in general still offers a lot of challenges. On the side of scheduled maintenance, the main inquiries are principally related to components' remaining useful life estimation, firstly to optimise scheduled maintenance planning, but also to try to move these maintenance activities toward condition based maintenance and make them therefore more transparent for operations by avoiding long immobilizations of the aircraft.

Unscheduled maintenance is still a big issue which is difficult to avoid, if not impossible, given the complexity of the aircraft themselves and of their operational environment. Therefore airlines, either by themselves or with the support of maintenance release & overhaul (MRO) organisations, have to set up an intricate and costly maintenance environment to make sure that, in the case of an unscheduled event, they will be able to quickly take action and ensure minimum disruption of operations, and this wherever their aircraft operate. That means that they may have to manage different main bases for heavy maintenance activities and outstations capable of handling most common unscheduled maintenance situations. This is in addition to spare parts provisioning, and management of tools, equipment, and skilled people required for the maintenance execution.

8.2.3 Managing unscheduled events

By definition, an unscheduled event can occur at any time, whether the aircraft is flying or not. While in-flight, aircraft core systems and functions are monitored by the flight warning system, which notifies the pilots whenever a malfunction has been detected. Additionally the pilots themselves can sense or notice something unfamiliar and require further analysis, or people on the ground can notice something unexpected while working around the aircraft. Based on this information, the different stages in the decision process involve assessing the impact of the defect on the aircraft operations and deciding on the maintenance to be performed in regard to operational and financial impacts.

8.2.3.1 Operational impact assessment

Following the detection of a defect, cf. Figure 8.5), pilots and airlines' operation controllers have first to determine what the impact of the event is on aircraft operations.

In the case of any critical event detected when the aircraft is flying, passengers and crew safety being the primary concern, the aircraft will be landed at the closest available airport compatible with the current aircraft state without any other considerations. If the event is not critical, then the outcome will depend on how the aircraft design can bear such a defect. This information is available in the aircraft minimum equipment list (MEL) and component deviation list (CDL) documents, which provide the list of equipment and components identified by the aircraft manufacturer as being required for a safe flight, and how failure of one of these elements affects this capability.

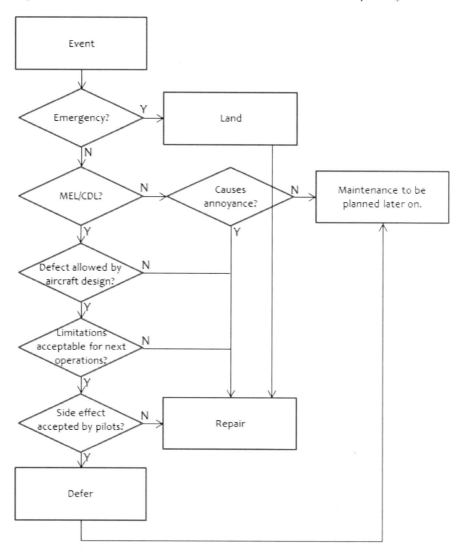

FIGURE 8.5
Event processing.

If the defect is stated in the MEL or CDL, then the decision will depend on

- Aircraft design accepting the loss of the component,

- Induced operational constraints that may affect the possibility for the aircraft to operate its next flights (for instance, if a flight level limitation applies whereas the aircraft will have to cross mountains),

- Induced additional burden for the pilots (having to periodically perform some visual inspections on some information to compensate the absence of a piece of equipment, for example) which the pilots may accept or not.

Additionally there may also exist some secondary technical events which may not be stated in the MEL or CDL but which can lead to some annoyance for the passengers or the flight crew (a flashing light, etc.) and that may also need to be repaired.

8.2.3.2 Maintenance activities identification

When the assessment of the operational impacts is available, the maintenance controllers or operators shall then determine what the requirements are at the maintenance level and how to manage them. According to the defect's impact and as stated earlier, in this context maintenance activity can either consist of deferring the defect or making a repair or replacement if the defect has an impact, or doing nothing and planning the repair later on if the defect is secondary. Deferring the repair is generally the quickest solution as it basically consists of setting the component in an inoperative state by securing it (locking a valve open for instance) or switching off its power supply. The repair or replacement activity may be more complicated, as it may require first isolating the root cause of the problem, which may be the component or equipment itself or another component or equipment interfaced with it such as wiring or a sensor, for instance, before being able to perform the repair or replacement. Additionally some tasks may involve specifically qualified maintenance personnel, as well as appropriate tools or ground supporting equipment (ladders, etc.), and potentially also spare parts or material.

8.2.3.3 Operational decision

Once the maintenance and operations controls has a clear view of the situation, they have to balance different options, taking into account the disturbances caused to passengers and other flights (delays, connections, cancellations,etc.) and the related financial impacts. For instance, they can decide to perform maintenance at the destination airport and transport required spare parts or missing equipment there if they are not available using another flight. They could also decide to move the aircraft to a location where it would be more appropriate to perform the maintenance. And to lessen impact on passengers, they could bring in a spare aircraft to continue operations, or swap the aircraft with another one where both will fit the next flights' operational requirements.

8.2.3.4   Maintenance preparation and execution

Finally, when the decision has been made, the maintenance personnel can prepare the required resources and logistical aspects so that maintenance can be performed with the least delays. Then once the aircraft has arrived, the maintenance activities – inspections, trouble shooting, repairs or replacements, software downloads, tests, etc. – can be started. These operations can be quite straightforward with simply pushing a button, or complicated with the need to access narrow areas, dismount panels to look for some components, access the cockpit to perform some tests, etc.

After completion, the maintenance operator reports the activity in the aircraft technical logs and a certificate of release to service is issued for the aircraft after a control of the operation.

8.2.4   Introduction to the aircraft maintenance system

The whole process described in the previous section, from the operational assessment to the maintenance decision, preparation, and execution, is generally highly time constrained with a lot of information to manage. And even though it can be thrilling to face such a situation and work to find out a solution, it also creates a lot of stress, as large amounts of money can be at stake. The purpose of the aircraft maintenance system is to help relieving this stress by providing new tools and new ways of working build around connectivity between the different parties and information digitalisation.

The aircraft maintenance system will allow people to share an overview of the operational and technical situation as well as the related operational impacts by integrating information from aircraft, flight operations, and maintenance systems with digitalised aircraft documentation. It will also provide support in maintenance preparation and maintenance execution by enabling troubleshooting activities, before the aircraft has landed, offering advanced connectivity with the aircraft during maintenance execution, and even automating some maintenance activities.

8.2.5   Aircraft maintenance system — System of Systems

8.2.5.1   Overview

The aircraft maintenance system architecture is built around two main concepts: the IoT, with the Arrowhead Framework, and service-oriented architecture (SOA). The IoT technology is used primarily to integrate aircraft, maintenance devices and the aircraft maintenance system. SOA will be used to interconnect the aircraft maintenance system with the other information systems involved in the flight operations and maintenance decision processes.

The aircraft maintenance system's System of Systems topology is depicted in Figure 8.6.

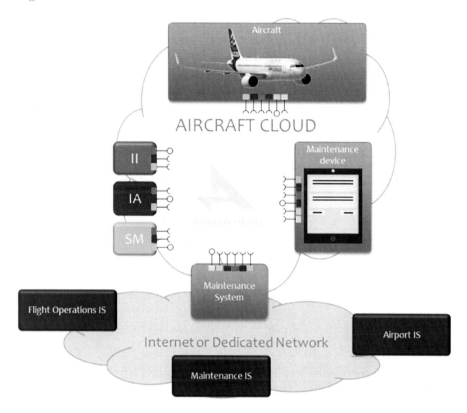

FIGURE 8.6
Aircraft maintenance system overview.

8.2.5.2 The Arrowhead aircraft cloud

The aircraft maintenance system is connected to a dedicated Arrowhead local cloud - that is a network with a dedicated Arrowhead Framework core services instance - altogether with aircraft and maintenance devices. This Arrowhead aircraft cloud allows the publication and use of services within a secure environment, restricted either to one airline and its partners, or to a set of airlines and maintenance organisations. The use of the Arrowhead Framework provides transparent integration of new aircraft and devices into the system, making it scalable, as well as built-in security and authorization management.

The different resources within the Arrowhead aircraft cloud instance follow a specific naming convention based on airline ownership, system category (aircraft, mobile device, etc.), and system identifier. This principle enables the setup of multi-tenancy mechanisms to support confidentiality and privacy of information and services between the different organisations sharing the cloud instance, and it also provides means to perform scale operations on a set of systems.

As an example, for a mobile device with identifier K43 managed by an airline which identifier is ABC, services will be named such as "ABC.MD.K43_notifyWorkOrder_ws_https._tcp." It is possible to look-up all the resources pertaining to the airline by requesting all the services prefixed by "ABC." to the ServiceDiscovery core service, or only the maintenance devices by requesting all the services prefixed by "ABC.MD.".

A similar principle is applied to the declaration of services by the aircraft maintenance system, supporting the management and isolation of the services contracted by the airline.

In addition the aircraft cloud makes use of certificates in combination with the Arrowhead Authorization core service allowing proper authentication of the different elements and access restriction to services.

8.2.5.3   Information system integration

On the other end, the aircraft maintenance system will be interfaced with other information systems (IS) providing structuring data and services using a standard SOA approach:

- The flight operations IS provides information regarding crew roster and flight schedules

- The maintenance IS provides services to look-up maintenance personnel skills and roster; hangar facilities, spare parts, tools and ground support equipment availability; and to allow booking these elements

- The airport IS provides information on airport facilities (gates, line stations, etc.) and resources (ground supporting equipments, etc.) availability.

8.2.5.4   Aircraft Maintenance System services overview

The airline maintenance system bridges the Arrowhead aircraft cloud to the airlines and partnering companies' information system and builds on these two environments to provide services to the end users.

Some examples of services that will be provided are

- "Dispatch impact assessment," which provides quick identification of the impacts and limitations induced on an aircraft by a defect and can be extended to provide information on impacts of the defect on the airlines' flight schedule

- "In-flight troubleshooting," allowing ground maintenance operators to start requesting information from the aircraft to get some insight on what lies behind a reported event, and thus allowing better preparation for the maintenance activities to be realized and saving time for their execution

- "Remote maintenance," which allows remotely performing some maintenance activities, with the possibility of automating some of them, and therefore can avoid flight delays or cancellation when no maintenance personnel are available at the destination to perform a deferral for instance

- "Mobile maintenance," allowing a mobile device to be interfaced with the aircraft and therefore simplifying interactions with the aircraft systems to perform some maintenance operations

8.2.6 Arrowhead aircraft cloud systems and services insight

As mentioned earlier, the Arrowhead aircraft cloud environment allows federating services for three categories of systems: aircraft, maintenance devices, and the aircraft maintenance system.

The following sections provide the definition of these systems and give an insight into some of the services they publish.

8.2.6.1 Aircraft system

The aircraft system basically represents an aircraft in itself. It provides an abstraction of the complexity of the aircraft by offering centralized and integrating access to information and services supplied on-board by the different equipment and components of the aircraft. Table 8.1 presents a sample of the services provided by the aircraft system.

8.2.6.2 Aircraft Maintenance System system

The aircraft maintenance system represents the gateway between the aircraft maintenance system and the Arrowhead aircraft cloud. The purpose of this system is to allow collection of data from aircraft and maintenance devices connected to the Arrowhead aircraft cloud, as well as enabling the publication of information toward these systems, or the requesting of some information. Table 8.2 presents some of the services provided by the aircraft maintenance system.

8.2.6.3 MaintenanceDevice system

The MaintenanceDevice system represents the device and applications that will be used by maintenance operators as a support for maintenance preparation, execution, and reporting. It consumes services provided by both the aircraft system and the aircraft maintenance system to render the different functions provided to the user such as the interactive job cab, remote support or the system pages. Table 8.3 presents some of the services provided by the MaintenanceDevice system.

TABLE 8.1

Aircraft services.

Service	Description
connect(<credentials>)	Allows a client system to request access to the aircraft services. Call to this service is compulsory for a client system to be authorised to use other aircraft services.
disconnect()	Revokes the access rights granted to the client system.
getConfiguration()	Returns the configuration of the aircraft.
runTest(TID,TestParams)	Runs a test on the aircraft, providing the given parameters, and returns the result of the test execution.
runCommand(CID,CmdParams)	Requests the aircraft to run the command identified by the provided command identifier with the provided parameters.
getParameters(P1,...,Pn)	Returns the current values for the list of given aircraft parameters.
recordParameters(P1,...,Pn,dur)	Requests the aircraft to record some parameters for a given duration. Returns the recording identifier and starts the recording of the parameters.
getRecording(RID)	Requests the recording corresponding to the given recording identifier. If the recording is still in progress, progress status information only will be provided.

TABLE 8.2

Aircraft Maintenance system services

Service	Description
notifyEvent(ACID, EvtData)	Allows a aircraft to notifying an event to the system.
notifyMaintenancePhase(ACID, Location)	Allows an aircraft to indicate it has entered the phase where maintenance is allowed, and confirms its location. The aircraft maintenance system can then notify maintenance devices they can start connecting to the aircraft.
registerDevice(DID, Location)	Allows a maintenance device to register itself into the system with information on its location.
isDeviceRegistered(DID)	Allows determining if a maintenance device is registered into the system. This is used by the aircraft to accept maintenance device connection.
unregisterDevice(DID)	Allows a maintenance device to unregister from the system.
isUserAuthorised(UID)	Allows determining if a user is registered in the system and can access the aircraft.

TABLE 8.3

Maintenance Device services

Service	Description
notifyWorkOrder(location, WO)	Allows the aircraft maintenance system to forward a work order to an aircraft mechanics located at a specific location.
getUserInfo()	Returns information on the user currently connected on the device.
notifyMaintenancePhase(ACID, location)	Allows indicating that an aircraft has entered the phase where maintenance is allowed, and confirms its location.

8.2.7   Use cases

The following section detail some use cases displaying the way the different systems operate within the Arrowhead aircraft cloud.

8.2.7.1   Automated maintenance scenario

Automated maintenance is a specific case of remote maintenance backed up by an automated procedure. It can be used to defer the treatment of a defect, for instance cf. Figure 8.7).

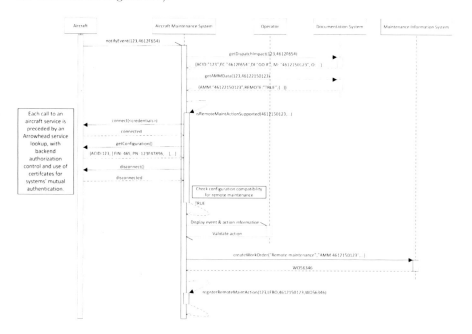

FIGURE 8.7
Automated maintenance decision sequence diagram.

After the aircraft notified the aircraft maintenance system, of the defect, the latter determines the related dispatch impact by referring to the aircraft MEL. The service returns information on the dispatch impact - GO IF in our example, which means that the aircraft can be dispatched under certain conditions - as well as the requirements for the aircraft dispatch - here a maintenance action to defer the repair by making the failed equipment inoperative.

The aircraft maintenance system then launches the process to determine if the action can be performed remotely. It connects to the aircraft and requests the aircraft configuration in order to validate that the versions of the software and hardware present on the aircraft are in line with the prerequisites for the remote maintenance action to be performed. If this is validated, then the aircraft maintenance system will notify maintenance control of the event and propose the execution of an automated remote maintenance action. Once validated by the user, the maintenance action will be planned for when the aircraft arrives, and recorded into the airline's maintenance information system.

When the aircraft has arrived at its destination and is ready for maintenance, it will notify the aircraft maintenance system (refer to Figure 8.8). This notification received, the aircraft maintenance system will again connect and authenticate to the aircraft and start the maintenance procedure: First it will confirm the presence of the failure by running a confirmation test. Then, the failure being confirmed, it will request the circuit breaker that powers the failed equipment to be opened by running the specific command. And finally, a new test is run to validate that the equipment has effectively been put into an inoperative state. This achieved, the aircraft maintenance system will record the maintenance action into the airline maintenance information system and into the aircraft's electronic logbook. Finally, the aircraft maintenance system will announce that the aircraft can be released to service and the aircraft will be able to depart once ground operations are finished.

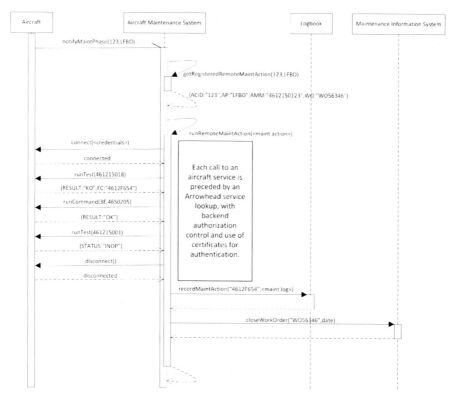

FIGURE 8.8
Aircraft automated remote maintenance execution sequence diagram.

8.2.7.2   Mobile maintenance scenario

Mobile maintenance devices provide digital support to the maintenance operator, allowing the interaction with both the aircraft maintenance system and the aircraft, therefore streamlining exchanges between the maintenance control and the operator on the one side, and on the other side making the operator more efficient by limiting needs to go here and there around the aircraft to access some specific functions or information. Additionally, the mobile maintenance device can provide some advanced interactivity with the aircraft that is not discussed in this scenario.

Whereas the operator connects to the mobile device, the device registers to the aircraft maintenance system together with user information and its location, and becomes visible from maintenance control. Following the detection of a defect on an aircraft and the decision to perform some maintenance to repair the defect, maintenance control will send an electronic work order to the maintenance operator. The work order provides the pre-requisites and description of the activity to be performed. Before the aircraft lands, the operator can prepare for maintenance by reviewing the documentation for the different tasks to be realised and by getting the resources and material needed for the operation.

When the aircraft arrives and is ready to accept maintenance operations, it notifies the aircraft maintenance system that it entered the maintenance phase, thus authorising mobile devices to connect to the aircraft (cf. Figure 8.9). The operator at that point requires access to the aircraft, which will validate that the device and the user are authorised to perform this action by referring to the aircraft maintenance system, and once validated the operator gets access to the aircraft. At that instant the device requests the aircraft configuration to validate which functionality can be enabled or not according to the aircraft's software and hardware versions. The operator then starts the trouble shooting activity with a fault confirmation test to validate that the defect is still present. The fault being confirmed, the operator will be guided step by step in the trouble shooting process by the system after the prerequisites of each step have been validated. In the example above, for instance, after having checked that the computer providing the failed function is working correctly, it will check that the fluid supply has been cut off before allowing the user to perform the next task. Once the fluid supply cut off, it will provide the operator with instructions to test a new valve in place of the current one, which will prove to solve the issue and indicate that the previous valve was the root cause of the defect. Then the operator performs a test to validate that everything is in order and can close the work order and submit valve replacement information to the aircraft's electronic logbook. The maintenance activity being done, the release to service of the aircraft can be pronounced and the aircraft will be able to depart when ground operations are completed.

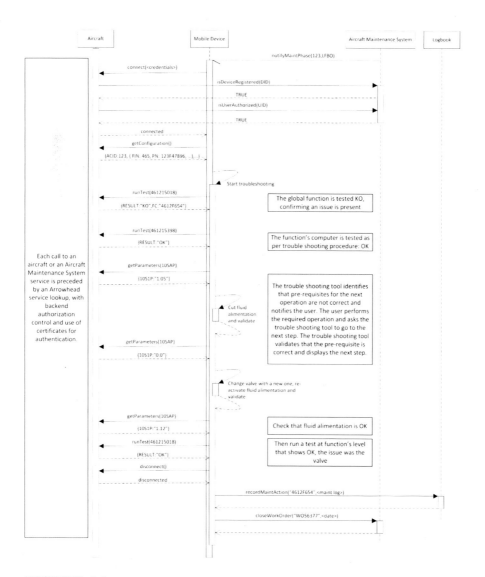

FIGURE 8.9
Mobile maintenance execution sequence diagram.

8.2.8 Conclusion

IoT principles and the Arrowhead Framework provide the means to easily set up scalable and secure systems such as the aircraft maintenance system requires to provide high-quality services in the face of a rapidly growing fleet of aircraft and higher customer expectations. The Arrowhead Framework offers the key components that make deployment and integration of systems such as new aircraft or maintenance devices within the aircraft maintenance system transparent, and enable focus on functions and information exchanges in an area where digitalisation has become the cornerstone of efficiency.

8.3 Glossary

CMMS: Computerised Maintenance Management System

DB: Database

DNS: Domain Name System

ERP: Enterprise Resource Planning

mDNS: Multicast Domain Name System

MEMS: Microelectromechanical systems

MIMOSA: Machinery Information Management Open Systems Alliance

MODBUS-TCP: Modbus RTU protocol with a TCP interface

MQTT: Message Queuing Telemetry Transport (protocol)

REST API: Representational State Transfer service interface

SWAGGER: Open API Initiative specification

OPC UA: OPC Unified Architecture (protocol and information model)

XML: Extensible Markup Language

Bibliography

[1] J. Laurila, A. Koistinen, E. K. Juuso, and T. Liedes, "Monitoring of a rod mill using advanced feature extraction methods," in *12th International Conference on Condition Monitoring and Machinery Failure Prevention Technologies 2015, CM 2015 and MFPT 2015*, 2015, pp. 580–590.

[2] A. Mathew, K. Bever, M. Purser, and L. Ma, "Bringing the MIMOSA OSA-EAI into an object-oriented world," in *Engineering Asset Management and Infrastructure Sustainability*. Springer Science + Business Media, 2012, pp. 633–646. [Online]. Available: http://dx.doi.org/10.1007/978-0-85729-493-7_49

[3] "OPC unified architecture (OPC UA) new opportunities of system integration and information modelling in automation systems," in *9th IEEE International Conference on Industrial Informatics, INDIN 2011*. Institute of Electrical & Electronics Engineers (IEEE), July 2011. [Online]. Available: http://dx.doi.org/10.1109/indin.2011.6035016

[4] A. Koistinen, "On-site calculations of generalised norms for maintenance and operational monitoring," in *Maintenance, Condition Monitoring and Diagnostics & Maintenance Performance Measurement and Management, MCMD 2015 and MPMM 2015*, 2015, pp. 104–109.

[5] D. Hästbacka, P. Aarnio, V. Vyatkin, and S. Kuikka, "Empowering industrial maintenance personnel with situationally relevant information using semantics and context reasoning," in *Proceedings of the 7th International Joint Conference on Knowledge Discovery, Knowledge Engineering and Knowledge Management*, 2015, pp. 182–192.

[6] E. K. Juuso and M. Ruusunen, "Fatigue prediction with intelligent stress indices based on torque measurements in a rolling mill," in *10th International Conference on Condition Monitoring and Machinery Failure Prevention Technologies 2013, CM 2013 and MFPT 2013*, vol. 1, 2013, pp. 460–471.

[7] E. K. Juuso, "Recursive data analysis and modelling in prognostics," in *12th International Conference on Condition Monitoring and Machinery Failure Prevention Technologies 2015, CM 2015 and MFPT 2015*, 2015, pp. 560–567.

[8] E. K. Juuso and D. Galar, *Current Trends in Reliability, Availability, Maintainability and Safety: An Industry Perspective*. Cham: Springer International Publishing, 2016, ch. Intelligent Real-Time Risk Analysis for Machines and Process Devices, pp. 229–240. [Online]. Available: http://dx.doi.org/10.1007/978-3-319-23597-4_17

[9] "The CMMS Benchmarking System," in *Reliable Maintenance Planning Estimating, and Scheduling*. Elsevier BV, 2015, pp. 439–452. [Online]. Available: http://dx.doi.org/10.1016/b978-0-12-397042-8.15006-5

[10] MIMOSA. Mimosa organization. [Online]. Available: http://www.mimosa.org

[11] Wikipedia, "Node.js — wikipedia, the free encyclopedia," 2016, [Online; accessed 25-October-2016]. [Online]. Available: https://en.wikipedia.org/w/index.php?title=Node.js&oldid=746166794

[12] D. Hästbacka, L. Barna, M. Karaila, Y. Liang, P. Tuominen, and S. Kuikka, "Device status information service architecture for condition monitoring using opc ua," in *Emerging Technology and Factory Automation (ETFA), 2014 IEEE.* IEEE, 2014, pp. 1–7.

[13] R. T. Fielding, "REST: Architectural styles and the design of network based software architectures," Doctoral dissertation, University of California, Irvine, 2000. [Online]. Available: \url{http://www.ics.uci.edu/~fielding/pubs/dissertation/top.htm}

[14] "Node-red." [Online]. Available: https://nodered.org

9

Application system design: Complex systems management and automation

Michele Ornato

CRF

Tullio Salmon Cinotti

University of Bologna

Alberto Borghetti

University of Bologna

Paolo Azzoni

Eurotech

Alfredo D'Elia

University of Bologna

Fabio Viola

University of Bologna

Federico Montori

University of Bologna

Riccardo Venanzi

University of Bologna

CONTENTS

9.2.1 Infrastructure design, development, and implementation

During the design and development, phases the adoption of a unified, open, and interoperable framework simplifies the design of the architecture of the charging infrastructure, both in terms of offered functionalities/services and in terms of ICT solutions, from the field up to the cloud. The electro-mobility solution based on the Arrowhead Framework, adopts a pervasive IoT framework (Eclipse Kura [2]) that is responsible for the edge operations: it abstracts and isolates the developer from the complexity of the hardware and the networking subsystems, and redefines the development and re-usability of integrated hardware and software solutions. Kura simplifies the design and development of the edge computing part of the charging infrastructure and contributes abstracting and hiding the technical details of the services exposed through the AF. On the cloud side, in turn, it simplifies the design and development of the core logic of the electro-mobility application, the use of existing services from third parties, and the creation of new business logic for specific electro-mobility applications and services.

The presence of a pervasive IoT framework on the charging stations simplifies the deployment of the whole charging infrastructure, providing provisioning functionalities and full remote control during the installation process.

9.2.1.1 Maintenance and optimisation

The Arrowhead Framework introduces a significant reduction of maintenance costs: the remote control of charging stations based on cloud and IoT technologies enables continuous monitoring, remote, predictive and planned maintenance, reduction of technical interventions on the field, optimisation of maintenance procedures, etc.

The Arrowhead Framework also allows the optimisation of the usage of the charging infrastructure. The capability to collect near real-time data from the charging infrastructure represents a huge opportunity in terms of analytics. The study of this huge amount of information allows the optimisation of the use of the charging infrastructure in terms of electric load balancing, costs for recharges, number of users served, and optimisation of the territorial coverage.

9.2.1.2 Monitoring

Another important source of cost reduction is related to the electricity distribution network that is required by the charging infrastructure. The availability of autonomous monitoring stations capable to recharge their internal batteries using alternative and green energy sources (i.e., solar panels) can significantly reduce the required territorial coverage of the electricity network, in particular in rural areas where only low power lines are available and the costs for improving grid infrastructure are higher. The charging stations based on the

Arrowhead Framework are fully autonomous also from an ICT point of view, allowing them to seamlessly become part of the global charging infrastructure and services.

The Arrowhead Framework IoT-based solution enables the creation of services that allow the connection of different electro-mobility players, providing a friendly access to charging stations. These services offer a solution for creating a provider-independent network of electric vehicle charging stations, a unified multivendor charging infrastructure (intercharge). A similar interchange platform could operate like roaming between different network operators on the international mobile communications market: the final user has one single, unified, simple way to access a variety of different charging stations spread over the territory. This is a significant example of the potential of the Arrowhead Framework in terms of **new service creation and new business opportunities exploitation**. There is, however, one obstacle to electric vehicle diffusion: the so-called range anxiety. Although all EVs have a driving range suitable for most users, this does not totally fulfil the common driver's expectation. Except for some high-end vehicles, EVs' range is lower than traditional vehicles and recharge time is significantly longer than gasoline/diesel refuel time. This is widely perceived as a lower usability of EVs in comparison to traditional vehicles. Arrowhead Framework electro-mobility infrastructure aims to overcome such obstacles by deploying a pervasive recharge network, characterised by the widespread availability of autonomous and remotely controlled charging stations. The pervasive coverage of the recharge infrastructure, especially in rural areas, is a crucial factor to compensate for the limitations of vehicle range, thus eliminating an important obstacle that will characterise electro-mobility for several years to come.

The adoption of a service-oriented solution also introduces indirect positive impacts. An important obstacle for electric vehicle diffusion is the cost of the batteries that, currently, has a significant impact on the vehicle price, with respect to equivalent petrol or diesel models. The 30%–40% of the value added in purely electric vehicles is still due to the batteries. The availability of an electro-mobility infrastructure that will encourage the purchase of electric vehicle, is directly linked to their mass production and the cost reduction is a consequence of mass production.

9.3 Electro mobility Systems of Systems

Electro mobility Systems of Systems (EM SoSs) are complex SOSs composed by heterogeneous entities that interact, cooperate, and interoperate to provide global services for electro-mobility, including electric vehicle recharge booking and management, charging infrastructure monitoring and maintenance, route planning [3], and EM services oriented to analytics.

EM SoSs are based on the Arrowhead Framework for the publication of EM services and for the creation of multi-domain application based on services published by third parties.

From a system architecture point of view, EM SoSs are composed of four macro-components that will be described in detail in the following sections:

- The electric vehicle and the recharge infrastructure;

- The Arrowhead SOA electro-mobility solution;

- The Arrowhead Framework EM services;

- The modular co-simulation platform.

The proposed EM solution focuses on three use cases:

- Vehicle recharge "on the move";

- Vehicle recharge in a private environment;

- Recharge stations in rural environments.

9.4   Electric vehicles and recharge infrastructure

Electric vehicles and charging stations represent the main subsystems of the distributed EM electrical infrastructure.

Within the vehicle, the battery, the vehicle main control unit, and, depending on the vehicle, a power converter are the components involved in the recharge process; therefore, they are the vehicle subsystems that interact with the rest of the recharge infrastructure. The amount of energy stored in a battery system can be very different from vehicle to vehicle, according to the vehicle purpose, performance, range, and battery characteristics. The battery's general behaviour can be described by the variable state of charge (SOC), an estimated value that spans from 0% (battery fully depleted) to 100% (battery fully charged). Furthermore, every battery system has its own operating limits, in terms of maximum and minimum operating voltage and operating current, that depend strongly on each battery's characteristics. These limits can never be overcome in any condition.

Recharge is mainly managed by a power converter that connects the energy grid (typically AC) and the vehicle battery being charged (a DC storage system). Such a converter is thus mainly an AC/DC converter that can be located either on board a vehicle or off board, depending on its characteristics and power capabilities. From the electrical point of view, the recharge sequence can be roughly described as a sequence of two phases:

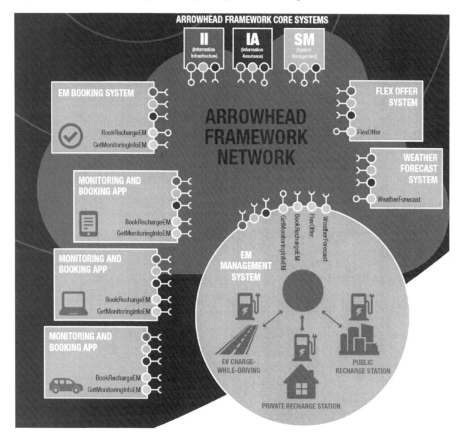

FIGURE 9.1
EM SoS overview.

- The first phase allows charging the battery from any starting level up to 70–80% SOC level, and is the constant current phase, in which the charger acts roughly as a current generator. The current value is defined by the charger capabilities (power) and the battery acceptance capabilities. The maximum voltage level is defined by the battery characteristics. During this phase, the battery voltage increases up to this maximum level. When this voltage level is reached, then the charger continues with the second phase.

- The second phase, which is needed to reach the full charge level, is the constant voltage phase, in which the charger acts as a voltage generator. The current is regulated (i.e., decreased) in order to keep the target voltage level without exceeding it. This phase ends when the current needed to reach the target voltage level decreases to 0 (i.e., full charge condition).

Of course, the charge sequence, from the electrical point of view, is not so straightforward and needs an information exchange between the battery and the charger: the details about this process are out of the scope of this chapter. In addition to the sequence previously described, it is worth remarking that a charge sequence has to be performed "as a whole", which means it cannot be interrupted but should not be segmented, i.e. once it has been interrupted it should not be restarted.

A typical charge profile, regardless of current or voltage values, is depicted in Figure 9.2.

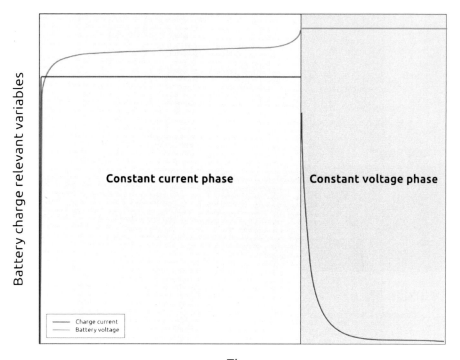

FIGURE 9.2
Typical recharge profile.

It must be highlighted that this kind of profile is applicable to every kind of battery, regardless of the charging technology or charging mode, as described in various international standards. It is a physical/chemical behaviour that cannot be modified.

From a simplified point of view, the battery voltage can be described as the sum of two contributions: one depending on SOC (the higher the SOC, the higher the battery voltage) and one depending on the current (coherent with current sign, so increasing voltage while charging and decreasing voltage while discharging). So far, the higher the charging current, the quicker the voltage limit is reached, thus preventing achieving a higher SOC level with that current. For a higher SOC level, current has to be decreased, thus slowing down the fast charging process. This means that a full charge can only be achieved by decreasing the charging current as soon as the maximum voltage is achieved. This means also that the charging sequence can be shortened only by increasing the charging current during the constant current phase. So far, fast charging will not be able to achieve full charge of battery, but only some 75-.85% SoC. Furthermore, within these SOC values, the higher the charging current, the lower the SOC level achievable.

The technical aspects of the charging process provide the requirements for the identification of significant use cases that represent the state of the art of charging technologies. Starting from these premises, in the Arrowhead project three different recharge use cases have been identified:

- AC slow charge for domestic electric vehicle supply equipment (EVSE)

- DC fast charge for public EVSE in remote locations

- contactless charge on-the-move.

These use cases are part of one of the project pilots, because each of them corresponds to a physical pilot showing the usage of the three different recharge technologies. The three recharge scenarios are not intended to be comprehensive, covering all the possible charging scenarios or technologies, as well as it is not intended to go through all the standardisation efforts currently running in Europe and worldwide about EV charging scenarios.

AC slow charge for domestic EVSE is the most common charge type. It is available to every electric plug-in vehicle on the market. The battery charger is quite compact, is limited to 16A or 32A (at grid level), and it is hosted onboard. The charging time is thus quite long, i.e., several hours, and the EV charge is typically intended to be performed overnight, to take advantage of off-peak low energy rates.

DC fast charge for public EVSE in remote locations (i.e., rural areas) is a particular case of DC fast charging that involves a new generation of battery charger that is capable performing a fast DC charge also in situations (remote locations) in which grid capabilities don't allow it. Such overperformance is possible by adopting a battery storage system used to accumulate the energy needed for a fast charge. Such a storage system can be charged either from the power grid or from a photovoltaic system also present on site. The battery storage system accumulates energy while no vehicle is charging and then will be quickly discharged during the vehicle recharge process. Of course, charging the auxiliary battery system will take a much longer time than discharging, due to grid or photo-voltic (PV) intrinsic capabilities, so the fast charge will be available only few times a day, according to the battery storage system capacity and to the total amount of energy requested by the incoming vehicles. Some additional considerations regarding fast charging technology are

- fast charging technology is available only for properly equipped EVs;

- the power converter for fast charging is typically a huge, high-power AC/DC converter located off vehicle, permanently connected to a power grid

- the charging interface to a vehicle is in DC.

The contactless charge on-the-move is a brand new charging technology that exploits jointly two novel concepts. The first is the contactless recharge, which adopts wireless power transfer technology to establish a power connection between one or more transmitting coil(s) located in the road, under the tarmac surface, and one receiving coil located in the lower part of the vehicle. This coil-based system transfers energy from the grid to the vehicle's battery without a wired connection. The second technology is the application of the coil-based approach to a moving vehicle, thus having a series of coils in the road, being activated and deactivated in sequence as soon as the vehicle being charged transits from one coil to the next one.

The three recharge use cases have been studied in the context of the Arrowhead project, and three distinct but interoperable pilots have been designed and implemented. The pilots are based and natively integrated within the Arrowhead Framework by means of Arrowhead EM services, which are the main topic of the current chapter.

Starting from the vehicle batteries and the charging stations, the automations of the EM complex system will be achieved through an end-to-end IoT cloud-based solution, an ICT infrastructure of hardware and software components that exploits the state of the art of pervasive systems, telemetry, cloud computing, and service-oriented platforms.

9.5   Arrowhead SOA EM solution

The core of the EM SoSs is a service-oriented end-to-end automation solution based on the Arrowhead Framework, on IoT technologies, and on cloud computing. The proposed solution has been successfully applied to EM real contexts identified in the three main use cases: private recharge, public recharge, and recharge on the move. The evaluation of the systems and of the application services, implemented to address the use cases, demonstrated that the required interoperability level among the hardware subsystems, the software pervasive subsystems, the services, and the cloud has been obtained. The analysis also showed that the implemented systems and services, by cooperating together, originate a tangible added value perceivable by both stakeholders and by final users, a positive indicator for the general acceptance and interest for EM.

The defined EM automation software architecture may hire an unbounded number of interacting services and systems, as it has been defined to be extendible; thus only the most relevant considerations about system interaction in the defined use cases of interest will be analysed in detail.

Figure 9.3 shows the main components of the EM automation infrastructure from an high-level perspective. The EM cloud platform constitutes the core of the architecture where the information from all the data sources is stored and shared with the other actors and services, according to privacy and information access policies enforced by the cloud management system. Sensible information to be managed in the cloud may come from different sources, i.e., electric vehicle onboard equipment, private or public buildings, energy storage systems based on renewables, charging stations, power grid components, or meters.

The cloud platform publishes the EM-Management services on the Arrowhead Framework and uses the services provided by the framework (i.e., the EM-Booking service, the WeatherForecasting service, the FlexOffer service, etc.). With this approach, the technical details are completely hidden and the functionalities of the EM SoSs are efficiently made available to other subsystems of the EM SoSs or to other cross domain applications. At the same time, the EM SoSs, interacting through the cloud with the AF, can exploit the services provided by other domains.

The charging stations represent the edge of the EM SoSs and exploit IoT technologies to interact with the cloud platform that, in turns, "converts" their functionalities and collected data in Arrowhead services. The charging stations adopt the Eclipse Kura IoT framework [2] for a pervasive integration with the cloud platform. Kura runs on a control unit inside the charging station that, acting as a multiservice gateway, provides the following functionalities:

- A hardware abstraction layer that simplifies the business logic development

- Wide support for data collection from the field

- Edge computing services for local data processing

- Efficient and secure cloud client that supports MQTT-based telemetry [4]

- Remote management functionalities that provide full remote control through the entire life cycle of the charging station.

The same pervasive IoT solution has been adopted for all the use cases.

The clients of the envisioned EM infrastructure may be fixed workstations, mobile applications or on board computers providing services on the move. The services may be schematised as entities with a client side, a server side, and a knowledge base on which they act. All these entities need precise information about the other subsystems in order to interact (i.e., addresses, access rights, information models, protocols, synchronisation policies, etc.).

The process of adapting the existing heterogeneous and multivendor components in order to properly interact and cooperate in common scenarios is difficult, expensive and implies many non-trivial challenges. In a similar complex scenario, the Arrowhead Framework represents the technological glue that connects together the data sources, the cloud infrastructure, the client equipment, and the services by requiring only simple elements. That is service publication is bound only to the requirements imposed by the authorisation service (in order to be trusted) and by the service registry (in order to be discoverable and to discover).

The application of the Arrowhead Framework to the EM scenario provides many advantages, among which is worth mentioning the separation of concerns through which each entity is given the opportunity of sharing only the data required to achieve service interoperability. All the other collected information remain completely hidden. As it is possible to see in Figure 9.3, the architecture of the EM infrastructure has been designed to be extendible with new services.

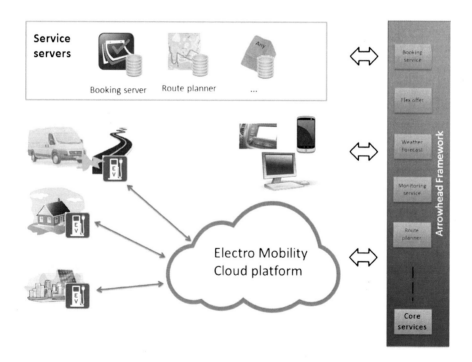

FIGURE 9.3
EM scenario: Automation infrastructure (high-level perspective).

9.6   Systems and services

The possibility to share system characteristics, features, and functionalities through open and interoperable service represents one of the key advantages of the Arrowhead Framework. It enables the simple creation of cross domain service-based application that fully exploits the potentialities of complex System of Systems. With this philosophy, the EM scenario becomes an ecosystem where multi-vendor and multi-domain solutions cooperate to provide new added value services to heterogeneous users and stakeholders.

The artefacts introduced in the EM scenario to create this ecosystem are (Figure 9.4) the electro mobility management system (EMMS), a set of services specific of the EM domain and a set of services from other application domains. The systems and services that have been involved are

- EM management system

- Booking service

- Route planner service

- Weather forecast service

- FlexOffer service [5]

The EMMS is the Electro Mobility System of Systems and is responsible for managing and monitoring the charging infrastructure deployed on the field, to manage its interaction with the cloud and to provide the EM service to the Arrowhead Framework. The services from the other application domain cooperate with the EMMS to create the added value application, that have been investigated and demonstrated in the EM use cases. More specifically, the actors and the entities involved in the use cases, through the Arrowhead Framework, are those defined in the three recharge scenarios illustrated in Section 9.3.

The first example of cross domain service usage is the booking service. The importance and the value of the booking system can be evaluated from three different perspectives.

From the user point of view, the booking service allows booking a recharge directly from a mobile device or from the charging station, at home or almost everywhere, also in rural areas, and this feature contributes to the reduction of driver anxiety, one of the biggest obstacles preventing the medium user from choosing an electric vehicle. Due to the long recharging time, it is paramount for users to be able to plan long travels considering the various recharge options.

From the charging infrastructure point of view, situations in which users crowd specific charging spots, leaving others underused, will give way to a more balanced and sound resource usage as the customers will be distributed on the available spots. Furthermore, this service allows collecting data that will enable the charging station owners to estimate the future demand and properly manage the local resources to face future usage trends. The booking service provides also an estimation of the minimum number of vehicles that are going to use the charging station, and this information is particularly relevant in the public recharge scenario, with local energy storage system. The public charging station, in fact, has to select the most convenient way to recharge the local storage; two options are available: exploiting solar energy with no added cost or relying on the power grid with the consequent costs, a faster method that is mandatory if there is the risk of completely depleting the local storage.

A third perspective under which the booking service may improve the EM scenario is that of the energy provider. The impact of EM on the electric distribution grid is relevant, as demonstrated by several scientific research papers [6, 7]. The power grid has different working modes: depending on the total demand it may range from high efficiency with low demand to low efficiency and high costs with high demand. In this context, the data collected by the booking service from the charging stations allows the energy distributor to improve the power demand estimations and so the power grid component usage, the costs, the quality of service, etc.

As anticipated, other services may improve the EM use case, if properly immersed in the AF ecosystem. The weather forecast service can be used in the public recharge scenario to improve the estimation of how much energy it is possible to store through solar panels in the coming hours. A route planner service may advise routes with wireless power transfer equipment in order to augment the probability of completing long trips without the risk of total battery depletion. The EM-Management system provides monitoring and management services that can be exploited at two different levels: directly in the electro mobility scenario and/or by other applications in other domains. The basic role of the EMMS is the orchestration of the distributed charging infrastructure: the main benefit is a shared solution based on IoT and cloud technologies for the integration of every single use case in the electro mobility scenario. This benefit is available at the scenario level and in other domains through EM services. For example, EM services are fundamental for the booking service in order to avoid the reservation of a charging station that is out of service or of a charging station on which a maintenance intervention has been planned. Two important areas of interest for the potential creation of cross domain applications are maintenance and multivendor intercharge. Maintenance activities can have greatly benefit using the EM-Monitoring service, because the remote control of charging stations simplifies their life cycle management, reducing the deployment and maintenance costs with provisioning functionalities, remote monitoring, remote management, and preventive maintenance. Intercharge is another example of cross domain application: EM services provide a common and shared interface that could connect various e-mobility players and provide EV drivers with simple and customer-friendly access to heterogeneous charging infrastructures. Different charging stations managed by different companies become part of the same network, and the new System of Systems does not only hide the complexity caused by this heterogeneity but offers to the drivers the same interface to receive real-time information about the status of the charging station or the charging process. Finally, the FlexOffer service may be used in the private recharge scenario to propose to the end user the best energy cost, thanks to a proper allocation of the effective recharging time, in a time interval where the power grid is not congested or, simply, the contract offers lower prices.

It is worth noticing that the contribution of the services to the quality level of the target scenarios, that is, the concrete added value, is achievable only if the exchanged information is trusted, if the services are able to seamlessly find and use each other's functionalities, and if a good grade of orchestration and cooperation among the services is supposed: in short, if the Arrowhead Framework core services are correctly used to manage the interaction among actors. For example, the booking service only allows booking of a charging sta-

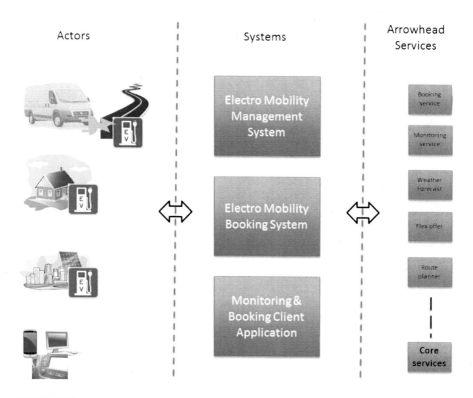

FIGURE 9.4

Automation systems and application services in the Arrowhead project for the EM scenario.

tions if the charge consumption data can be received by charging spot owners or energy distributors. Moreover, the quality of service offered will be lower if the booking service can't use the knowledge base of the monitoring service to improve its reliability, e.g., to avoid reserving a charging spot currently in maintenance or to properly advise the end user and propose other recharge options when a fault happens on a previously reserved charging spot. Similar considerations are valid for the route planner and also for the other services involved in the electro mobility scenario. The route planner, indeed, needs updated information about the new charging station installed on the territory, while the energy distribution company needs the data of the route planner to derive statistics from the user needs and find optimal positions to install new charging stations or charging stations on the move.

Each service involved in the electro mobility scenario is provided by a specific subsystem; therefore, there are no specific security issues due to service interactions: the data shared will be only those exposed through the Arrowhead Framework and the interacting entities will communicate using messages encrypted on the basis of security certificates validated by the Arrowhead Framework Authorization service.

Figure 9.4 shows that the integration in the Arrowhead big picture is not

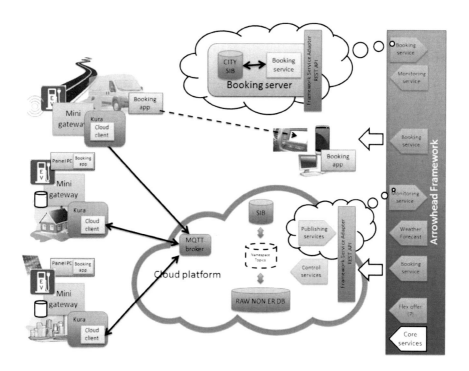

FIGURE 9.5
EM scenario: Interoperability components and software localisation.

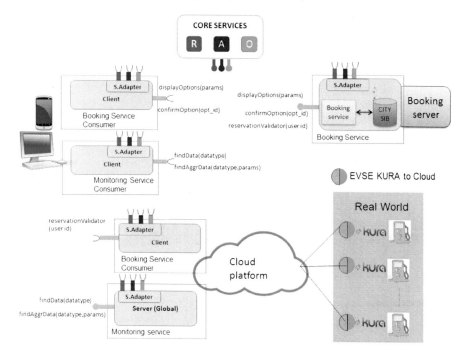

FIGURE 9.6
Booking and monitoring application systems.

invasive as only a REST interface has to be provided for the subsystems to cooperate. In regard to the cloud, the main integration components are the publishing and control services, defining which data to share and with whom. The information coming from the charging infrastructure is collected adopting a device-to-cloud approach based on the Eclipse Kura framework and, with the MQTT telemetry protocol, is sent to and stored in the cloud. The raw data are made interoperable and machine understandable through a semantic annotation process using a semantic information broker (SIB) to manage the resulting semantic assertions.

Smart-M3 [8, 9] has been adopted as a middleware for interoperability. It is a semantic publish-subscribe data-driven interoperability platform, which stores data encoded as Resource Description Framework (RDF) triples with respect to ontological rules. The data stores are called semantic information brokers (SIBs) [10, 11] and provide APIs in order to perform operations such as insert, update, delete, query, and subscribe, all of them encoded with the SPARQL [12, 13] or RDF-M3 formalisms.

Analysing and comparing Figures 9.5 and 9.6, it is possible to understand how the integration allows two of the main EM subsystems to cooperate. The booking server was a standalone legacy service that has been adapted to the framework through a properly designed REST interface. This interface can be used by both end users wanting to reserve a charging station and the charging station equipment to check the user identity. This second interaction starts from a control unit installed on the charging station which starts an information exchange through the Kura-based device to the cloud platform; the request triggers a cloud internal control service which relies on the Arrowhead Framework to discover the correct instance of the booking service and to check the user credentials. A failure of this check prevents the reservation of the recharge process from starting.

The cloud publishing services are used when information about entities connected to the cloud is needed by external agents. The cloud internal logic manages the calls to the findAggrData primitive, a function used by the EM-Booking service (after discovering the EM-Management service) to detect malfunctioning charging stations in a given area. A reply is provided only if the caller has the right to access such data.

9.7 Arrowhead EM services and related automation aspects

This section describes the arrowhead services designed and implemented in the electro-mobility scenario. The description is based on the main sequence and class UML diagrams specified in the documentation artefacts SD and IDD, cf Section 3.3[14]. These documents, together with the CP and SP document, cf Section 3.3[14], provide the knowledge base for service interoperability and for their usage in the Arrowhead Framework.

9.7.1 EM-Monitoring service

The EM-Monitoring service provides two primitives for the interaction with the charging infrastructure (see Figure 9.7): *findData* and *findAggrData*. The *findData* primitive is used when an EM subsystem or a third party application needs information about a specific charging station. Using this primitive, a client is able to query the monitoring service about the whole status of an EVSE by simply specifying its identifier in the *findData* primitive.

The *findAggrData* primitive, instead, is used when the EM agent needs information about a group of charging stations sharing some property, or when it is necessary to refer to a charging station indirectly, through metadata, instead of directly through an identifier. This primitive is particularly relevant because it provides information that is useful in the most common EM use cases. This simple interface generates a rich set of information that corresponds to conceptually complex queries: for example, calling the *findAggrData* primitive it is possible to identify all the charging spots in a circular area ge-

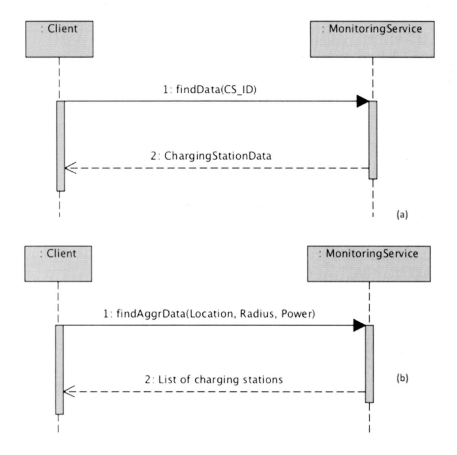

FIGURE 9.7
EM-Management system: Information retrieval from a single (a) or multiple charging stations.

ographically centered on a specific point with a specific radius, identify the closest charging stations where it is possible to recharge 40 kWh in less than 3 hours, or simply detect if a fast charging station is currently in operation.

The information model of the EM-Monitoring service is illustrated in Figure 9.8. The model explains how to query the EMMS and describes which information can be retrieved from a charging station with the EM-Monitoring service.

9.7.2 EM-Booking service

The sequence diagrams in Figures 9.9, 9.10, and 9.11, show the main operations performed by the EM-Booking system: the *updateknowledgebase* loop

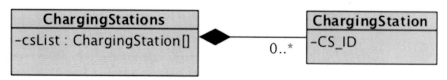

FIGURE 9.8
EM-Management system: Information model.

represents the interaction between the EM Booking system and the EM-Management system to allow the latter to rely on an updated list of EVSEs and to properly inform users about recent events regarding their recharge reservations.

The booking application service operates according to a handshake strategy between the client and the server module. First, the client asks for possible booking options by specifying its preferences in terms of space, time, and amount of energy needed. Then the service, according to its business logic and to the requirements specified by the user, provides a list of charging options. At this point the user may choose one of the provided options, finalising the reservation, or make a different request.

The sequence diagram in Figure 9.11 shows the service behaviour when a conflict happens in the same time frame. It is not uncommon that, by scaling the end user number, reservation conflicts due to overlapping reservation on the same EVSE at the same time could represent a concrete issue. In this case

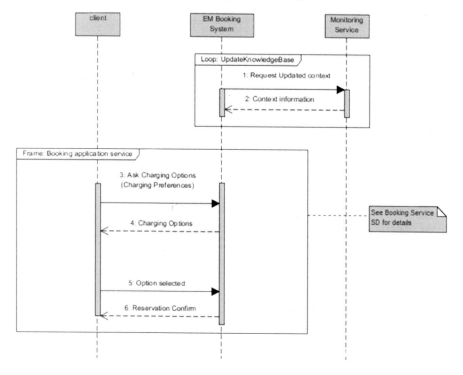

FIGURE 9.9
EM-Booking service: Booking protocol and knowledge base update.

the user which confirms the last the reservation is informed that the booking option is not available anymore due to a concurrent booking from another user and is invited to choose another option or make a new query to the booking service. The information model shown in Figure 9.12 illustrates the main data structure used by the booking service to answer user requests, verify user identity, interact with the EM management system, and, in general, perform all the operations necessary to properly manage the EVSE booking.

9.8 Co-simulation platform

Energy and smart cities scenarios are considered among the most interesting and powerful environments in which automation systems and services prove their usefulness. However, such systems are often tricky to demonstrate, since the required architecture, especially the hardware components, is not in

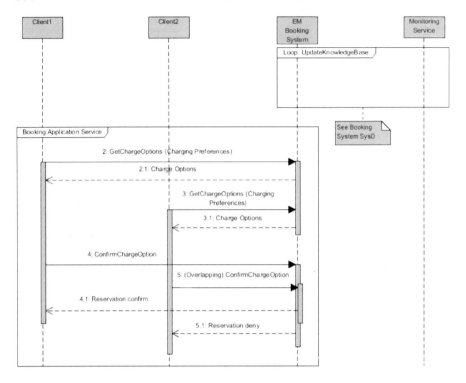

FIGURE 9.10
EM-Booking service: Authorisation check.

place. This also motivates simulations as both validation environments and pre-deployment analysis tools.

The University of Bologna developed a co-simulation framework capable of depicting an urban scenario with vehicular traffic, electro mobility, charging stations, and smart grid model [15]. In particular, the developed simulator allows the analysis of urban traffic, including a significant percentage of electric vehicles and clusters of charging stations (in particular, fast charging stations concentrated in public parking lots). The co-simulator platform described in this section has been created in order to meet the following goals:

- Design a distributed control able to mitigate the congestions in the network caused by an excess power demand by the recharging of electric vehicles;

- Test the infrastructure from the services' point of view, both as a pre-deployment analysis and a scalability and benchmark test;

- Give to the developer the possibility of building automation systems on top of a sandbox which resembles a realistic scenario.

This section provides a glimpse of the architecture of the co-simulation

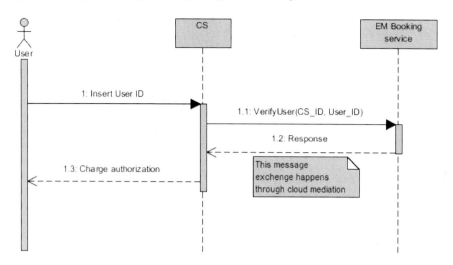

FIGURE 9.11
EM-Booking service: Information model.

platform and of its components as well as its Arrowhead services, illustrates the analysis carried out by using the simulator through the presentation of some results, and depicts the possibilities for automation systems and services operating within such an ecosystem.

9.8.1 Architecture of the co-simulation platform

The co-simulation platform integrates the traffic simulator and the power distribution simulator. Figure 9.13 illustrates the general architecture of the co-simulation environment.

Smart-M3 has been adopted as a middleware for interoperability. In such a scenario, the city SIB collects information regarding all the entities coming from different domains, while the dash SIB collects data about the vehicles.

9.8.1.1 Traffic simulator

Urban traffic is modeled using VeinS, which is an open source framework for vehicular network simulations based, in turn, on two simulators, namely, discrete event-based simulator OMNeT++ and road traffic simulator SUMO. SUMO (Simulator of Urban Mobility) is an open source traffic simulator capable of modeling entities such as roads, vehicles, traffic lights and vehicle routing. Each entity is simulated microscopically; thus it is possible to interact with them separately. OMNeT++ is a general purpose simulation environment for communications, which is able to model customisable and interoperable modules. A large-scale scenario (i.e., downtown Bologna) was

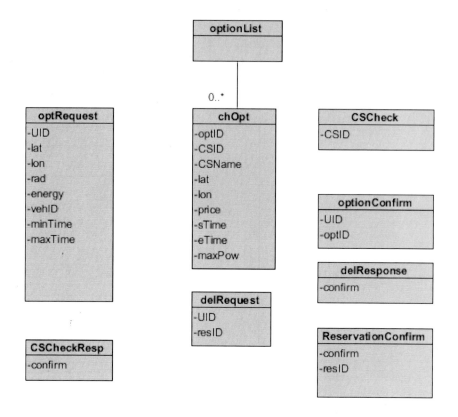

FIGURE 9.12

EM-Booking service: Information model.

considered, with a realistic street map (imported from the OpenStreetMap project). VeinS has been extended with the models of EVs and EVSE units (including the management of the EVs queues) and it has been integrated with the battery charging/discharging models described in [16]. In particular, OMNeT is used to characterise the electric vehicles through the simulation of the battery charging and discharging process, the charging stations, and the previously described reservation mechanism.

Depending on the simulated time of the day, a different traffic rate is established (i.e., total number of vehicles running in the scenario at the same time). When an EV reaches a level of battery lower than a predefined threshold it tries to go to recharge. Results show that the use of the reservation mechanism implies significantly less recharging failures.

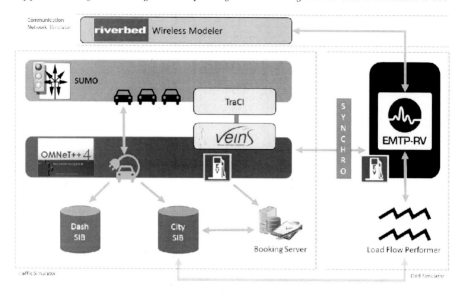

FIGURE 9.13

Architecture of the co-simulation environment.

9.8.1.2   Power distribution system simulator

The power distribution system simulator is based on an EMTP-rv (electro-magnetic transient program), which is a time domain simulator. The distribution feeders are modelled including models of the three-phase unbalanced lines, of the three-phase HV/MV substation transformers and of the aggregated unbalanced loads (constant impedance/current/power) that include the EVSE units. The model of the EVSE is based on a triplet of current sources. For each phase the control relies on a feedback regulator in order to inject or absorb the requested values of active and reactive per-phase power, as described in [17] and [18]. Each aggregate load is connected to the secondary side of a MV/LV transformer.

9.8.1.3   Arrowhead Framework services

The simulated scenario is intended to resemble the real scenario and the use cases previously defined. In particular, as illustrated in Figure 9.14, it implements an instance of the Arrowhead Framework EM-Management system (defined as a requirement for EM scenarios) and it is fully interoperable with the Arrowhead Framework EM-Booking system. Each vehicle in the simulation performs a call to the EM-Booking service whenever it needs to recharge and picks randomly one of the options displayed.

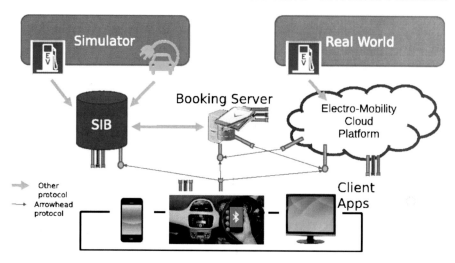

FIGURE 9.14
Mixed real-simulated scenario.

9.8.2   Mobile service platforms and significant results

In the context of service granularity, in particular when referring to electro-mobility, the world of mobile, in-vehicle, and context-aware services has a fundamental role: simplifying the transition to the new infrastructure by both reducing the impact on end users, who feel anxious about new technologies, and also encouraging the transition to EVs. Realistic EV scenarios are characterised by the heterogeneity of the actors involved in them, i.e., different EV models, EVSEs providers and smart-grid operators, each using their data representation techniques, and providing their own services, mobile applications, or APIs for third-party software deployment.

In such a chaotic environment the developed mobile services were run, completely unaware of the world that they are running in, meaning that they have the same behaviour when they are deployed in real-world entities or simulated entities. This happens because simulated vehicles can also be controlled remotely through an Arrowhead mobile application, which is indeed the same application that a real user would use while seated in his or her electric car, exploiting both the EM-Booking service and the EM-Monitoring service. Such services show the same interface to consumers regardless of the scenario in which they are deployed. The set of applications having such a property is called "Mobile Application Zoo (MAZ)".

As a result, a mobile application will be able to retrieve and display data of a simulated vehicle (e.g., current charge state), as if the user were driving it. Vice versa, interacting with the mobile application, the user might perform actions that trigger events within the simulation, for instance, a reservation might result in the simulated vehicle changing its route.

A MAZ component provides two main advantages. First, it allows testing the correct behaviour of mobile applications under several different conditions that might occur in a real-world scenario. Second, it allows evaluating the performance improvement provided by the utilisation of the mobile applications described in the following subsections when all the scenario components are in the loop, i.e., the vehicular traffic, the charging/discharging operations, the impact on the smart-grid.

9.8.2.1 EM-Booking service

As in the envisioned real scenario, the EM-Booking service uses the EM-Monitoring service in order to get geographical information about the EVSEs nearby and books a recharging time slot in a target area, and during a preferred time frame. In Figure 9.15 is shown screenshots from mobile application client using the reservation service. On the left is the result of a user query, i.e., all the EVSE that match the user preferences and are returned to the client app. When a user selects his preferred recharge option, the application shows the path from the current position to the EVSE and the time at which the reservation is made (Figure 9.15 right).

FIGURE 9.15
Reservation service.

The test of the reservation process within the simulated scenario gave the results shown in Figure 9.16, which depicts the average EVSE utilisation when using or not using the reservation service. When a reservation is not used, the simulated EVs seek for an available EVSE, driving around the city. Figure 9.16a shows that the average occupation increases with the number of electric vehicles. The average occupation without a reservation is generally higher than that with reservation because simulated EVs fill the nearest charging spots fast without any coordination. This chaotic behaviour brings battery depletion risks when the number of EVSE is low with respect to that of EVs needing to recharge in a certain zone. In Figure 9.16b, it can be noticed that the number of EVs that run out of battery without being able to recharge, is much larger when a reservation is not present. This happens because when all the charging spots are filled, the residual EVs, being unaware of the situation, continue to travel until the residual charge ends. The results provide two important indications for service developers and energy distributors. On one side, they clarify the usefulness of the reservation service, because the EVSEs are utilised in a coordinated manner, which translates into much more efficient scheduling of charging operations. On the other side, they are valid as a predeployment analysis, since they provide feedback about the optimal planning of the charging infrastructure, in terms of number and location of EVSEs.

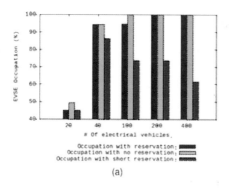
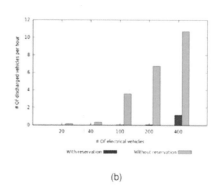

(a)   (b)

FIGURE 9.16

Evaluation of mobile services for EVs: EVSE occupation for reservation service.

9.8.2.2   RoutePlanning service

This section describes the RoutePlanning service, which is depicted in Figure 9.17 [19][3].

One of the most relevant issues of electro-mobility is the uncertainty of successfully traveling through a planned trip. This is due to the complexity of the accurate estimation of the consumption along the path and to the low

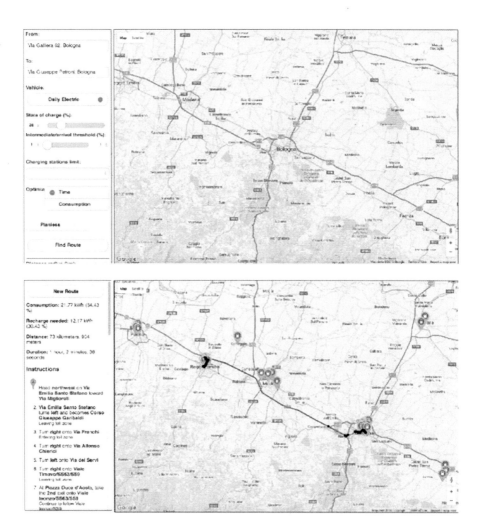

FIGURE 9.17
RoutePlanning service.

density of charging opportunities. Hence, a tool to assist the EV driver has been developed: it computes the expected consumption over the desired path, and identifies the needed charging opportunities by minimising either the total travel time or the total consumption.

The route planner uses EVSE data provided by the EM-Monitoring service and works as follows. First, it computes the expected consumption between the start and the end of the path. If a user-defined threshold of intermediate charge is satisfied, then the system simply returns the desired path with the directions to reach the destination. The threshold, called Intermediate State Of Charge (SOCint), was introduced as a safety threshold that users can tune based on their anxiety.

If the path is not feasible (due to a violation of the SOCint parameter), the algorithm searches for an available charging spot, minimising the deviation from the original path. In order to avoid the problem of looking for a charging opportunity farther and ending up in a longer trip, it looks for EVSE closer to the destination compared to the starting point. From each EVSE which can be reached without violating the SOCint threshold, it looks at all the paths to the destination, if feasible. If the algorithm finds one which does not exceed the SOCint parameter, it looks for a feasible path from the EVSE to the destination. Among all the feasible paths, it takes the one that either minimizes the consumption or the travel time, according to the user preference. If it cannot find a path with the previous step, then it again execute the algorithm to find an additional EVSE in which we can charge starting from the previous EVSE, which becomes the starting point for the next step. The algorithm repeats the previous steps until either it finds a feasible path and returns it to the user, or it cannot find any, and thus it returns to the user that it is not possible to travel through the desired path with the chosen parameters.

A large-scale scenario has been simulated, i.e., the Italian Emilia-Romagna region, taking into account real EVSE positions and 3D street maps. Random trips were generated within the scenario, and the success probability was studied, i.e., the probability of reaching the destination when following the indications of the route planning service or when using a conventional approach commonly adopted by EV drivers. This approach consists in following the shortest path, and in seeking for an available EVSE, only when the charge is below a given threshold. Figure 9.18 shows that the SP decreases (dashed lines) without route planning but also decreases, in a much slower way, with route planning. This happens because the coverage of EVSEs on the target scenario is not uniform: most of the EVSEs are located on urban areas, while charging opportunities are quite scarce in rural areas. However, as before, Figure 9.18 provides useful feedback for service planning, since it demonstrates the performance gain of a planning service for medium and long trips, and for grid planning, since it allows detecting the areas of the Emilia-Romagna region which are mostly uncovered by the current charging infrastructure.

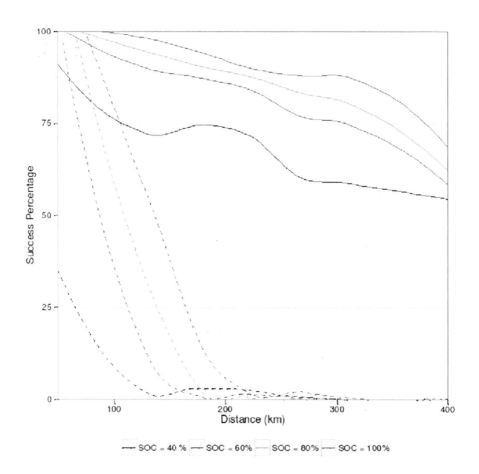

FIGURE 9.18
Success probability for the route planning service.

9.9   Conclusions

The automation of complex systems is a challenging context which has to deal with many factors, including technology readiness, business opportunities, user appreciation, legal regulations, heterogeneity of legacy systems and privacy issues. From this point of view, the electro-mobility scenario represents a relevant case study where an automation software architecture based on the Arrowhead Framework has been designed and implemented. The identified solution has been conceived to meet the scenario requirements, identified through the analysis of some relevant electro-mobility use cases selected in the Arrowhead project. The resulting software architecture is based on a device-to-cloud approach that efficiently exploits the IoT enabling technologies and the computational resources offered by the cloud infrastructure, providing an evident improvement in terms of interoperability, information security, and extendability. Among the heterogeneous set of systems and application services that is possible to plug in the architecture, the EM management system and the booking system have been analysed in detail, designed, and implemented. With respect to other services provided by the Arrowhead Framework, their importance for the electro-mobility scenario has been highlighted or envisioned with simple considerations: the selection included, but in principle is not limited to, WeatherForecast, FlexOffer and RoutePlanning services. One of the most important results inferred from the performed analysis is that the full potential of a service is available only when it is integrated in a whole automation infrastructure with the ability to interact with other services for the orchestration of new advanced functionalities.

The evaluation of the architecture and the services has been performed through a co-simulation framework taking into account the traffic, the variable percentage of electric vehicles, the power grid elements, and the implemented set of services. The simulation environment, according to the mobile application "zoo philosophy," allowed realistic tests with real services applied to a simulated power grid and a simulated set of electric vehicles and users. The obtained results confirmed the validity of the proposed automation infrastructure, showing a clear improvement of metrics like the load balance of the power grid and the percentage of long trips ended without total battery discharge.

Taking into account all the considerations done and the presented results, it is possible to confirm that the application of the Arrowhead Framework to complex scenarios allows their complete automation by improving the interaction between the subsystems and the cloud and preserving, at the same time, the indispensable security and reliability requirements.

Bibliography

[1] P. Varga and C. Hegedűs, "Service interaction through gateways for inter-cloud collaboration within the arrowhead framework," *5th IEEE WirelessVitae, Hyderabad, India*, 2015.

[2] "Eclipse kura documenation." [Online]. Available: http://eclipse.github.io/kura/

[3] S. Mehar, S. M. Senouci, and G. Rémy, "Ev-planning: Electric vehicle itinerary planning," in *Proc. International Conference on Smart Communications in Network Technologies (SaCoNeT) 2013*, vol. 01, June 2013, pp. 1–5.

[4] (2016) Mqtt is a machine-to-machine (m2m)/"internet of things" connectivity protocol. [Online]. Available: http://mqtt.org

[5] L. L. Ferreira, L. Siksnys, P. Pedersen, P. Stluka, C. Chrysoulas, T. le Guilly, M. Albano, A. Skou, C. Teixeira, and T. Pedersen, "Arrowhead compliant virtual market of energy," in *Proceedings of the 2014 IEEE Emerging Technology and Factory Automation (ETFA)*, Sept 2014, pp. 1–8.

[6] L. P. Fernandez, T. G. S. Roman, R. Cossent, C. M. Domingo, and P. Frias, "Assessment of the impact of plug-in electric vehicles on distribution networks," *IEEE Transactions on Power Systems*, vol. 26, no. 1, pp. 206–213, Feb 2011.

[7] K. Clement-Nyns, E. Haesen, and J. Driesen, "The impact of charging plug-in hybrid electric vehicles on a residential distribution grid," *IEEE Transactions on Power Systems*, vol. 25, no. 1, pp. 371–380, Feb 2010.

[8] J. Honkola, H. Laine, R. Brown, and O. Tyrkko, "Smart-m3 information sharing platform," in *Proc. IEEE Symposium on Computers and Communications.* IEEE, 2010, pp. 1041–1046.

[9] D. G. Korzun, A. M. Kashevnik, S. I. Balandin, and A. V. Smirnov, "The smart-m3 platform: Experience of smart space application development for internet of things," in *Internet of Things, Smart Spaces, and Next Generation Networks and Systems.* Springer, 2015, pp. 56–67.

[10] F. Morandi, L. Roffia, A. D'Elia, F. Vergari, and T. S. Cinotti, *RedSib: a Smart-M3 semantic information broker implementation.* SUAI, 2012.

[11] E. Ovaska, T. S. Cinotti, and A. Toninelli, "The design principles and practices of interoperable smart spaces," *Advanced Design Approaches to Emerging Software Systems: Principles, Methodology and Tools*, pp. 18–47, 2012.

[12] E. Prud'Hommeaux, A. Seaborne *et al.*, "Sparql query language for rdf," *W3C recommendation*, vol. 15, 2008.

[13] A. Seaborne, G. Manjunath, C. Bizer, J. Breslin, S. Das, I. Davis, S. Harris, K. Idehen, O. Corby, K. Kjernsmo *et al.*, "Sparql/update: A language for updating rdf graphs," *W3C member submission*, vol. 15, 2008.

[14] F. Blomstedt, L. L. Ferreira, M. Klisics, C. Chrysoulas, I. M. de Soria, B. Morin, A. Zabasta, J. Eliasson, M. Johansson, and P. Varga, "The arrowhead approach for soa application development and documentation," in *IECON 2014 - 40th Annual Conference of the IEEE Industrial Electronics Society*, Oct 2014, pp. 2631–2637.

[15] A. D'Elia, F. Viola, F. Montori, M. Di Felice, L. Bedogni, L. Bononi, A. Borghetti, P. Azzoni, P. Bellavista, D. Tarchi *et al.*, "Impact of interdisciplinary research on planning, running, and managing electromobility as a smart grid extension," *Access, IEEE*, vol. 3, pp. 2281–2305, 2015.

[16] L. Bedogni, L. Bononi, M. D. Felice, A. D'Elia, R. Mock, F. Montori, F. Morandi, L. Roffia, S. Rondelli, T. S. Cinotti, and F. Vergari, "An interoperable architecture for mobile smart services over the internet of energy," in *Proc. IEEE World of Wireless, Mobile and Multimedia Networks (WoWMoM) 2013*, June 2013, pp. 1–6.

[17] R. Bottura, A. Borghetti, F. Napolitano, and C. A. Nucci, "Ict-power co-simulation platform for the analysis of communication-based volt/var optimization in distribution feeders," in *Innovative Smart Grid Technologies Conference (ISGT), 2014 IEEE PES*, Feb 2014, pp. 1–5.

[18] R. Bottura and A. Borghetti, "Simulation of the volt/var control in distribution feeders by means of a networked multiagent system," *IEEE Transactions on Industrial Informatics*, vol. 10, no. 4, pp. 2340–2353, Nov 2014.

[19] L. Bedogni, L. Bononi, A. D'Elia, M. D. Felice, M. D. Nicola, and T. S. Cinotti, "Driving without anxiety: A route planner service with range prediction for the electric vehicles," in *2014 International Conference on Connected Vehicles and Expo (ICCVE)*, Nov 2014, pp. 199–206.

10

Application system design - High security

Andreas Aldrian
AVL List GmbH

Peter Priller
AVL List GmbH

Christoph Schmittner
Austrian Institute of Technology

Sandor Plosz
Austrian Institute of Technology

Markus Tauber
Fachhochschule Burgenland

Christina Wagner
Austrian Institute of Technology

Daniel Hein
IAIK TU-Graz

Thomas Ebner
Evolaris Next Level

Martin Maritsch
Evolaris Next Level

Thomas Ruprechter
Infineon Technologies AG

Christian Lesjak
Infineon Technologies AG

CONTENTS

10.1   Introduction

An important aspect of IoT automation systems is the global distribution of production environments. Such production environments are of huge financial value and hence need continuous maintenance, which is cost intensive due to the distributed nature of such systems. If, however, too much maintenance is carried out, resources are wasted and production processes are interrupted. If too little maintenance is carried out, the risk of system failure is increased. This chapter describes the investigation of an automotive industry use case. The equipment itself is designed for non-Internet scenarios; nevertheless there is the requirement for service optimisation which requires collection of data from the industrial device via the Internet. Such data is mission critical for the customer, and the industrial equipment must always be protected to provide the same level of operational safety and measurement result stability. Figure 10.1 shows that the data is transported across trust boundaries, i.e., via the Internet. Thus cyber risks exist which could have an effect on operational safety. Therefore the IoT automation system must include the following features:

(i) Authentication and certification

(ii) Reliable message subscriber systems

(iii) Being engineered with continuous assessment against safety and security threats

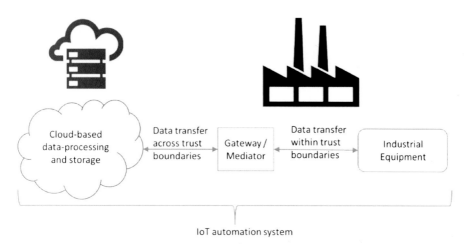

FIGURE 10.1
Basic architecture of industrial IoT automation systems.

In the course of the project the methods applied lead to the fact that it is required to utilize

a) a service that authenticates machines reliably

b) requires a firewall friendly protocol that is ISO/IEC 20922

c) that any architecture needs to be evaluated according to security and safety methods.

All methods that were applied and evaluated for use in the Arrowhead context and have been contributed to the Arrowhead Framework documented in detail here, respectively in Chapter 6.

The remainder of this chapter is as follows: in Section 10.2 we outline the use case in detail and in section 10.3 we describe how it has been implemented and how the authentication service was realised. In Section 10.4 we explain how the ISO/IEC 20922 protocol has been implemented, and in section 10.5 how safety and security assessment has been applied. In Section 10.7 we reflect again how ISO/IEC 20922 based services did help us to build a reliable highly secure system with the help of parts of the Arrowhead Framework.

10.2 Smart maintenance use case

The IoT automation system under consideration enables "Smart Service" applied to measurement devices in engine test beds used for verification in the

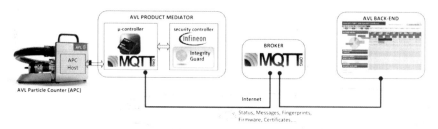

FIGURE 10.2
Smart service architecture.

automotive industry. It is going to be implemented by AVL List GmbH in a prototypical manner and discussed with partnering customers. It is intended to use collected data to provide smart services after a trial and settling-in period. In this first phase only a limited number of options will be available. However, first indications and trends can at least be deduced to estimate the impact of a full-featured solution.

A mediator (see architecture overview in Figure 10.2) connects the target system (e.g., a particle counter device APC) with

(i) a trusted, "internal" network and

(ii) the Internet.

The mediator communicates with the target system based on a serial interface that does not support any connection-based functions like routing. DoS attacks targeting the overall system coming from the Internet therefore can only affect the operation of the mediator but not the target device itself. As the mediator provides a hardware-based separation between external networks (Internet and trusted internal network) and the target system, configuration and traffic control on the network becomes much easier. For a safe, reliable, and secure implementation of the mediator, focus was put on strong separation, software diversity and complete input validation between the interfaces, to protect also against common-cause and cascading failures. Its hardware is based on a Sitara A8 or A9 system running Linux. Software and hardware prototypes have been implemented covering the following major requirements:

- physically separate the customer (trusted) network from the Internet

- a firewall compatible, "inside-out" initiating, lean protocol has been chosen to transfer device status information to the provider of smart services via the Internet (ISO/IEC 20922)

- end-to-end connections are fully protected by TLS encryption

The main technical requirements of the mediator concept are as follows:

- *unique ID*: it shall be possible to uniquely identify any device attached to a mediator

- *secured WLAN* (radius) connection for fast data transfer between a mobile device (e.g., service technician's tablet) and target systems

- *NFC*: for local operation the mediator shall offer a short- range communication interface for certificate exchange and to ensure that any operator that needs to get a link between the device and the Internet is physically close to the device itself

- *secured storage*: owners need to be able to define what a device is allowed to send to the smart service provider (or what it is not allowed). Those definitions shall be stored in a secure location which only owners have access to

- *storage*: it shall also be possible to store non-critical data such as documentations on the mediator

- *certificate store (PKI)*: a certificate store shall be supported; security updates shall be possible

- *licensing*: it shall be possible to control certain features of the target device via the manufacturer's on-line licensing system

- *AAA conformity*: a device shall be able to meet AAA conformity

- *VPN remote*: it shall be possible to establish a remote tunnel connection to a device

- *MQTT protocol*: to exchange information as lean as possible the mediator shall incorporate MQTT (ISO/IEC 20922) as the communication standard

- *crypto*: the mediator shall be able to encrypt and decrypt any information going in or out

- *clock*: to support security mechanisms a clock functionality shall be incorporated

- *policy updates*: as several customers have a large number of devices it shall be possible to control and update configuration (e.g. what a device shall send to the Smart Service provider) via centralised policy servers

- *adaptive GSM*: as an optional requirement a device shall be able to provide connectivity via cellular radio technologies such as EDGE, UMTS, LTE, etc.

Figure 10.3 illustrates a modularised approach of the proposed mediator design. Based on the security objectives given by a customer and the target environment, HW and SW components can be tailored for an optimal system. The current mediator design captures all requirements of the smart service industrial use case in a single, modular design.

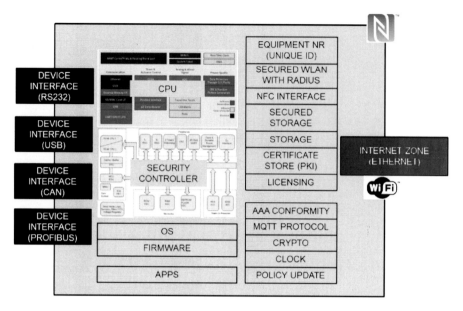

FIGURE 10.3
Mediator overview.

10.3   Authentication and certification service

As described in the previous section, information exchange between the smart service provider on one hand and the plurality of monitored devices (target systems) on the other hand is done via MQTT (ISO/IEC 20922), and further structured by location, type, owner, etc. Information like device status, logs, usage data. etc. called snapshots will be typically provided by the device by publishing it into a specific topic (category) on the MQTT broker. For obvious reasons, it is important, that only authorised clients are permitted to publish snapshots or to subscribe to topics.

First, device owners should not be able to publish snapshots of industrial equipment they do not own. Furthermore, they should not be able to subscribe to another owner's equipment snapshots. Finally, while the vendor and (or smart service provider should be able to collect all data, owners should only be able to read data for equipment they own. By subscribing to snapshots of an owner's own equipment, they can transparently monitor what data is acquired from their devices. In general, the mediator should be able to gather data from the attached industrial equipment and transmit it to one or more smart service providers. For this process, data end-to-end confidentiality and authenticity need to be ensured. Specifically, data must be protected while it is in transit between the mediator and the smart service back-end. The ISO/IEC 20922 based data distribution infrastructure requires an authorisation framework to ensure that system owners only access topics which are relevant to them. Specifically, they should only be able to push and pull messages for devices they own. Example: a device owner will be able to push operational data for any of her devices and pull firmware updates for those.

In order to provide a secure system and to fulfil the defined security requirements, the system uses three mechanisms:

- End-to-end messaging payloads are protected and encoded using Cryptography Message Syntax (CMS)

- Each communication link which leads through the Internet is secured with Transport Layer Security (TLS) encryption.

- The MQTT broker is equipped with a filter which only allows authorised entities to access topics and messages

10.3.1   Transport layer security

The smart service scenario relies on a Public Key Infrastructure (PKI) to distribute keys and establish connections through a trusted party (certificate authority (CA)). Entities (mediators, administrators, broker, backend) are identified by a certificate which binds a public key to their name and is signed by the (trusted) CA. The entity has to hold the corresponding private key for the certificate in order to decrypt messages intended for them. The private key must be stored in a secure environment. Certificates must be revoked and replaced, should a third party obtain access to a private key.

10.3.2   End-to-end encryption

For end-to-end messaging, the operational data is protected and encoded using the Cryptographic Message Syntax (CMS) standard. Operational data is encrypted and authenticated by the mediator using Authenticated Encryption (AE), specifically AES-GCM. Authenticated Encryption is a symmetric cryptography scheme that achieves data authenticity and confidentiality using a single ephemeral symmetric key. Here was used CMS's key encapsulation capabilities to wrap the symmetric key under an ephemeral wrapping key using the "One-Pass MQV, C(1e, 2s, ECC MQV) Scheme."

Figure 10.4 shows the creation, transmission, and unpacking of the CMS message. The left side shows the encryption of a plain text (coloured in blue) and all information, which is required to encrypt the text for particular recipients (identified by their certificates). The resulting CMS message (encoded in ASN.1) is indicated in the center. The encrypted text is now encapsulated in the *authEncryptedContentInfo* field (coloured in blue). On the recipient again, the information required to unwrap the key and decrypt the text is indicated as well as the results after processing the CMS message.

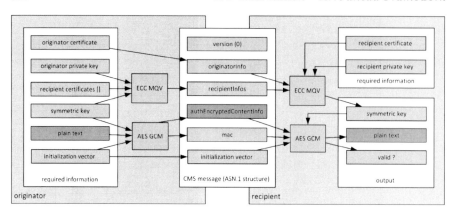

FIGURE 10.4
CMS structural overview.

10.3.3   Authorisation filter

For authentication/authorisation, the MQTT protocol specifies username and password fields for MQTT CONNECT messages. However, these credentials are encapsulated in plain text within the MQTT message and also sent in plain text to the MQTT Broker. Thus, an adversary who can eavesdrop the MQTT message or who is capable of accessing the broker, can obtain the credentials. In addition, using a username-password authentication would require the system maintainer to build and administer an extra authentication system. This plain-text transmission of username-password requires the use of transport encryption. The system already requires for all clients TLS connections with client authentication to connect to the broker at all. Therefore, all clients (equipment, vendor dashboard, customer dashboard, etc.) are already equipped with a client certificate to authenticate their identity. The filter should make use of these identities not only in the TLS client authentication but also in the topic authorisation process, i.e., it should restrict publishing and subscribing access to topics to certain publisher and subscriber groups, which will be determined based on the identity corroborated by every client's certificate and associated credentials. This broker-side security component is called Topic Access Control System (TACS). The TACS is able to grant or deny access to certain topics based on the client's identity. In a practical setup, an equipment's certificate is not directly issued by the smart maintenance service's Certificate Authority (CA), but by an intermediate CA of the customer operating the equipment. Thus, the TACS would also be able to verify accesses on customer topic level. The filter component uses the identity of the connecting customer device, which is proven by the TLS client certificate authentication process, to compute the authorisation of the device versus the data channels in the broker it protects. This is a logical mechanism, which filters all channels not designated for a given customer device. If a filter is compromised, for example, by a successful remote software attack, the attacker will be able to access all information available to the attached broker. The attacker might also be able to mount attacks against broker/filter combinations higher up in the chain. It must be assumed that the attacker has full access to all encrypted messages in the system. The filter is currently directly built into the Mosquitto MQTT Broker. Based on the *Common Name* of the client certificate (which every client has to present to the broker in order to prove their identity; see above), the broker will either grant or deny access on the client's desired topic.

10.4 ISO/IEC 20922 Support for the smart service architecture

MQTT which is standard ISO/IEC 20922 [1] is a binary publish-subscribe messaging protocol working on top of the Transmission Control Protocol/Internet Protocol (TCP/IP). There are three components in MQTT: message publishers, message subscribers, and brokers, that connect publishers and subscribers. Messages are exchanged with an associated topic, also known as subject or channel. An MQTT broker manages messages through hierarchical topics. Publisher clients publish messages into certain topics and subscriber clients are able to subscribe to subsets of the available topics they are interested in. The multilevel topic structure allows clients to efficiently subscribe to the desired topics only. When subscribing, wildcards may be used. A plus sign (+) represents a single-level wildcard and a number sign (#) represents a multilevel wildcard.

MQTT provides a multi-level topic hierarchy that is used in the target scenario when aggregating and exchanging snapshot data. Topics are both routing information and metadata. Thus, not only the MQTT message payload needs to be protected, but the potentially valuable or sensitive information in the topic information as well. Transport Layer Security (TLS) protects topic labels while in transport. However, we also want to ensure that only authorised clients are able to write into and read from topics which are intended for them. Clients need to actively publish messages to brokers. This is suitable for connecting industrial equipment to smart maintenance services, as a company's information security policy will likely prohibit that external servers pull data from the equipment. With MQTT, the equipment needs to actively initiate connections to the outside. A broker in the demilitarised zone (DMZ) instead of directly connecting into the backend on vendor side is suggested for multiple reasons: first, as the Broker is at a known public location in which all data comes together, we consider it the most vulnerable link in the chain, from customer devices to a vendor backend. Thus it is likely to be the primary point of attack for adversaries. Furthermore, it is generally of vendors' interest to avoid remote Internet connections into the backend. When using MQTT, there will be no connection actively established into the backend from the outside; connections to the broker need to be initiated by a subscribing backend component.

During the investigations, a proto-typical implementation of the backend was set up. For the MQTT broker the open-source message broker Mosquitto was applied and placed in a DMZ. The backend components at the smart service provider side were realised in Java, in which open-source Paho Java Client for data retrieval was utilised (i.e., MQTT subscriptions). An additional component decrypts the end-to-end encrypted data within the backend which is encrypted inside the mediator on the customer side. As data is arriving in different formats depending on the equipment, a data interpreter will parse and sanitise values from snapshots for further processing.

10.5 Safety and security analysis for identifying system vulnerabilities

The Austrian Institute of Technology (AIT) performed safety and security analysis of the different architecture generations for the Arrowhead pilot "Smart Service" by applying different methodologies within this context [2], [3]. Various methodologies have been consolidated over the years and their applicability for the given context has been evaluated. Suitable candidate methods have been provided to the Arrowhead Framework and documented in Chapter 6. Here we document a method which we have developed by combining existing security assessment methods. It is structured as depicted in Figure 10.5.

The first step defines security objectives which are to be considered during the whole assessment. The next step is to build a model of the system and identify possible threats. Hence, Microsoft's STRIDE methodology is used for this step to create a catalogue of the most relevant threats, which are then further evaluated to assess their impact, and likelihood of the risk involved. In order to support our analysis, as inputs for assessing encountered threats, interviews with system experts from the device vendor have been performed through questionnaires, surveys, or personal consultations. Further description of each step can be found in Chapter 6.

10.5.1 Threat model of the architecture of the pilot

The initial step when performing the threat modeling procedure is to create a detailed Data Flow Diagram (DFD) for a particular architecture in order to identify criticalities, trust boundaries, and potential threats.

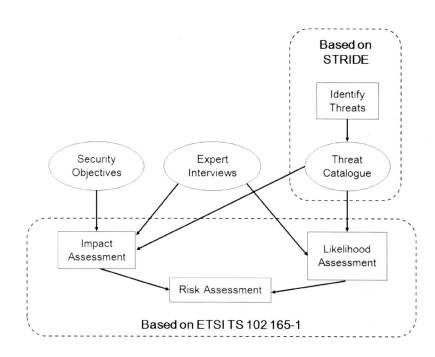

FIGURE 10.5
Security analysis methodology.

ISO20922 + HW security as enabler for secure inter-cloud communication

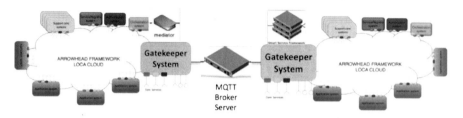

FIGURE 10.6
Usage of the MQTT end-to-end secure channel to implement the Arrowhead Framework Gatekeeper core system.

10.5.2　Interviews of system experts

During the interview process with system experts we got to know the system in more detail, allowing us to clarify some technical details and help us perform risk assessment more thoroughly and accurately. The output of the expert interviews is used to adjust the system threat analysis by identifying criticality points and sensitivity of information in the smart service pilot system architectures.

10.5.3　Risk analysis

In total 89 possible threats have been identified. These have been further analysed, and 20 of them were found directly applicable to the realised system. These have been put under risk assessment.

10.5.4　Safety analysis

Safety analysis methods like FMEA have been evaluated for their applicability in the bespoke domain. They have been contributed to the Arrowhead Framework and documented along with security analysis methods in Chapter 7 of this book.

10.6　Integration with Arrowhead Framework

The here described secure end-to-end transport mechanism is one of the two technologies identified to implement the Arrowhead Framework Gatekeeper system, see Section 3.4.2.8. The usage is depicted in Figure 10.6

10.7 Summary

Gaining attention largely due to the rise of distributed production environments, smart maintenance can be defined as a technique to predict when a device will fail so that maintenance can be planned in advance. Optimised maintenance minimises production downtime, use of resources, and thus costs; it ensures the reliability and in some contexts also the safety of the machine operation. Therefore, in this chapter we have described the investigation of a smart maintenance use case in the automotive industry. To ensure that only authorised clients are permitted to publish snapshots or to subscribe to the topics, we have implemented software and hardware platforms such as:

(i) Cryptographic Message Syntax (CMS) standard for end-to-end encryption

(ii) Transport Layer Security (TLS) for secure physical end-to-end connection

(iii) MQTT, an ISO/IEC 201922 standard, for authentication and authorization.

The Austrian Institute of Technology (AIT) has also developed a security analysis methodology, mainly consisting of security objectives derived from interviews with system experts, threat model and risk analysis.

The overall goal of the activities was to evaluate the above technologies and methodologies for an IoT automation use case, i.e., smart services.

Bibliography

[1] "ISO/IEC 20922 - Information technology – Message Queuing Telemetry Transport (MQTT) v3.1.1," ISO, Standard, February 2016.

[2] S. Plosz, M. Tauber, and P. Varga, "Information assurance system in the arrowhead project," *ERCIM News*, no. 97, p. 29, April 2014.

[3] S. Plosz, C. Lesjak, N. Pereira, M. Tauber, T. Ruprechter, and A. Farshad, "Security vulnerabilities and risks in industrial usage of wireless communication," in *Proceedings IEEE ETFA 2014*, Barcelona, Spain, September 2014.

11

Application system design - Smart production

Daniel Vera

FDS

Robert Harrison

Warwick University

Bilal Hamed

Warwick University

Claude le Pape

Schneider Electric

Chloe Desdouits

Schneider Electric

Hasan Derhamy

Luleå University of Technology

CONTENTS

11.1   Automotive manufacturing

11.1.1   Introduction

The automotive industry has become extremely competitive due to the increased market demand for frequently renewed and highly customised car products that must exhibit drastically high levels of quality. The current evolution of the automotive industry towards hybrid and fully electric vehicles is further positioning the sector in a potentially risk averse phase because of the introduction of new powertrain technologies and new associated manufacturing processes. Mass production of batteries, power packs, and in the future fuel cells, as well as electric motors and complex electrical components, represents a largely unknown territory for the current car manufacturing industry.

This phase of uncertainty will require an increased level of agility and responsiveness from the automotive manufacturing industry in order to integrate the constant changes and evolutions that occur in the design of electric powertrain components (e.g., new battery technology, evolving battery pack design, etc.), and also in order to achieve a ramp up from today's relatively low volume PEV (Plug-in Electric Vehicle) and PHEV (Plug-in Hybrid Electric Vehicle) production to the mass production that growing US, Chinese, and EU markets will require in near future. In particular, the manufacturing of electric powertrain components requires additional control over equipment and operator safety due to the handling of chemicals and high-energy-density material (i.e., batteries when charged), process quality control, process and product traceability as well as process flexibility (i.e., detection of faulty cells/packs and application of corrective processes, rework of faulty packs). Due to the high value and polluting nature of the material they contain, electric powertrain components will also require the instantiation of advanced reconditioning and recycling processes in order to adapt to a new product life cycle.

All the above require a radical increase in the ability to monitor and control the manufacturing processes very closely. This implies the use and deployment at a large scale of so-called Industrial Internet of Things (IIoT), smart devices, and connected equipment in order to implement and realise the benefits of the Cyber Physical System (CPS) as defined by the Industrie4.0 initiative [1, 2]. In this context the Arrowhead Framework provides a key element in achieving connectivity and integration between various layers of the manufacturing systems and organisations (i.e. from shop floor to top floors and across supply chain partners, see Chapter 1), and in facilitating the access to and management of data (cloud-based services related to data storage, processing, and analytics as well as data security) required to achieve the targeted objectives as defined above.

11.1.2 Automotive manufacturing: CPS system life cycle

In the domain of automotive manufacturing, the Arrowhead Framework (see Chapter 3) was considered through its link with the Industrie4.0 vision and the concept of CPS (see Chapter 1); The concept of CPS is typically used to describe a system composed of connected devices (IoT or IIoT) that have some forms of data or functional representation in a "cyber" (i.e., software/digital layer). In the context of manufacturing, there is an increasing number of devices and systems which can be categorised as IoT, IIoT, or CPS enabling devices [3]. To name a few: smart tools, sensors and conditions monitors (e.g., energy, temperature, vibration, position tracking systems) as well as vision systems, IP cameras, and also programable logic controller (PLC) nodes and more complex systems such as automatic guided vehicles (AGVs), industrial robots, or even human operators and manual operation cells equipped with smart sensors and/or mobile device interfaces.

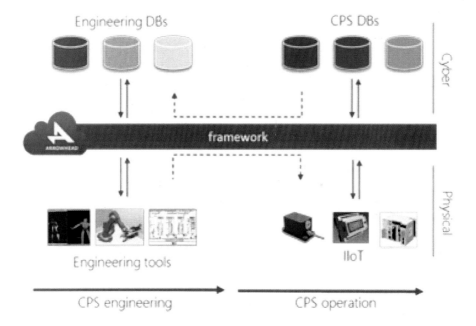

FIGURE 11.1
Cyber physical system engineering life cycle.

The concept of CPS and the use of connected/IoT devices in the shop floor represents a real opportunity for the manufacturing sector to significantly improve engineering processes, not only during operation of production systems' life cycles (e.g., monitoring, maintenance), but also across engineering projects and manufacturing programs. In the automotive industry, productions systems are reconfigured rather than redesigned from scratch and several car manufacturers use global line designs (e.g., Nissan, BMW) which evolve throughout various programs [1]. The data collected from IoT devices in the shop floor during line operations can be used not only to support MES, SCADA and ERP related functionalities (stakeholder collaboration domain as defined in Figure 1.2 of Chapter 8), but is also an invaluable source of information and knowledge (i.e., cyber signature of the physical system) that can be used to support system and production optimisation across programs (life cycle domain as defined in Figure 1.2 of Chapter 8) [3, 4].

In the context of the Arrowhead project the manufacturing consortium represented by FDS (software solution provider), HSSMI (non-profit research organisation), WMG (University of Warwick), and the end user Ford Motor Company, UK, has focused primarily on the use of the Arrowhead Framework to enhance support for operation monitoring and maintenance of production systems. The deployment of IoT devices and the CPS approach to management of automation has the potential to significantly improve the reactivity of maintenance operation for break-fix operations and also to facilitate preventive maintenance operations and management. The scenarios investigate how facilitated connectivity between condition monitors, PLC, and mobile device application can allow more effective shop floor level maintenance operations. Additional scenarios were investigated which include the deployment of automatically generated PLC code to PLC devices and management of PLC control code and configurations.

11.1.3   Maintenance of automation systems

Effective monitoring of production systems is critical in

1. detecting faults as soon as they occur in order to minimise the impact of those fault (e.g., downtime, but also impact on product quality and equipment integrity) and in

2. accurately assessing the state of the system in order to anticipate faults and enable more effective and accurate predictive maintenance. Another aspect of maintenance is

3. to enable rapid and effective intervention on the shop floor after a fault is identified or once a work order is scheduled

The time to intervention and time to repair are essential key performance indicators (KPIs) of maintenance operations efficiency. Non-productive intervention time includes time to identify faulty assets, time to locate assets in the shop floor, time to identify fault type and root cause(s) as well as time to access repair procedures and instructions and the required contextual information (e.g., electrical drawings, control cabinet layout, etc.).

The pilot used to implement the Arrowhead Framework based maintenance solution involves the use of a full scale automation machine located at

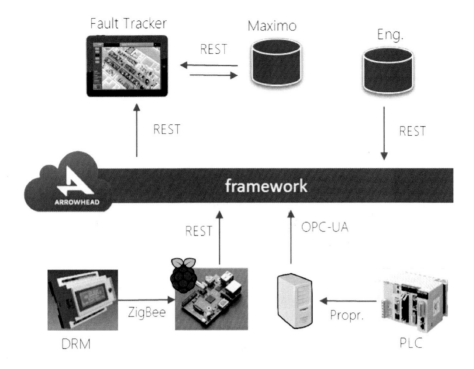

FIGURE 11.2
Data Sources for PLC automation maintenance scenario.

WMG, university of Warwick, referred to as the Automation System Workbench (ASW). The ASW is based around a conveyor loop and eight modular stations and is fitted with PLC based control systems from various vendors (i.e., Siemens, Schneider, Mitsubishi, Bosh Rexroth, and Allen-Bradley) and is configured to support the assembly of 18650 battery cells into battery packs. PLCs are connected to an OPC-UA [5] server (Kepware OPC-UA solution) using PLC vendor-specific drivers. So-called Dynamic Resource Monitors (DRMs) developed by HSSMI are installed on two of the ASW stations to monitor energy consumption and temperature at various locations in the system. The DRMs communicate with REST server implemented on a RasberryPi hardware via the ZigBee protocol.

In this configuration, the Arrowhead Framework is used to facilitate the orchestration between OPC-UA and the RasberryPi server that are providers of respectively control state and energy information, and the so-called fault tracker (FT) application developed by FDS. The FT application provides three sets of functions:

1. retrieve work orders from the MAXIMO maintenance database used by Ford. The MAXIMO database is linked to a MES system that provides a

list of assets that are in fault state, preventive maintenance schedule and work orders as well as maintenance history. The FT is also designed to

2. provide contextual information about the faulty asset; this include live system data such as current control state of all components in the system which are in fault, recent energy log, and also engineering information such as 2D plant layout, 3D model of faulty asset, PLC function block I/O mapping, electrical wiring diagrams and finally vendor or maintenance specific information such as instruction manuals and repair procedures, work steps, and safety and hazard information and procedures and finally

3. the FT application also allows the maintenance engineers to feed data (i.e. textual information, work steps, photos) back to the MAXIMO database. The orchestration between service providers (i.e., OPC-UA server, DRM monitors, engineering database) and the service consumers (i.e., Fault Tracker application) is provided by the Arrowhead Framework Orchestration services.

An overview of the communication sequence between the system's components is provided in the diagram in Figure 11.3.

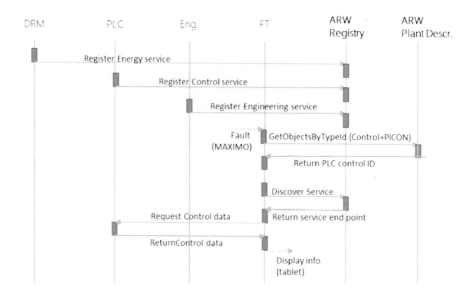

FIGURE 11.3
PLC automation maintenance scenario using the Arrowhead Framework and vueOne Fault Tracker application.

1. Service providers register their services using the Arrowhead ServiceRegistry (see Chapter 3).

2. The FT application retrieve Fault information from the MAXIMO database

3. The FT uses the PICON number from the MAXIMO database to interrogate the Arrowhead Framework PlantDescription system (see Section 3.4.2.1) and retrieve a list of all nodes of a certain type (e.g., control, 3D/VR, DRM) linked to the given PICON number. Note that the Plant Description Arrowhead Core service acts here as an ID translation service; different types of IDs are used to identify various devices which are part of the same system node. For instance, a station in the production line is identified in the MAXIMO database with a PICON number. The same station is controlled using a PLC that is identified using an OPC-UA server IP address and one or more PLC IP addresses, and one or several DRMs may be fitted to a station each identified by a unique ID. In addition, the same station is identified using a physical brass plate fixed onto the machine frame.

4. The FT then retrieves the end point for the service it requires (e.g,. energy, control or engineering data access) and use the relevant IDs to retrieve contextual information about the faulty device.

5. This information is presented to the maintenance engineers through different screens on the FT tablet application. For instance, faulty asset location is provided by concatenating 2D layout of the plant with area/zone/section information retrieved on the MAXIMO database. Similarly, a 3D model of the faulty asset is retrieved, allowing the operator to visually identify the faulty component. The current PLC control state can be retrieved and viewed along with the history of the energy monitors fitted on the station or the asset itself.

The above scenario provides an example of how an IoT communication framework, such as Arrowhead, can be used to achieve connectivity and integration between

1. IIoT devices and

2. other sources of data issued from engineering data bases and a

3. mobile device used on the shop floor and therefore provides an access point toward which both live CPS data (i.e. from connected devices) and engineering data (i.e., from various engineering databases) can converge.

11.1.4   PLC control code deployment

Current engineering methods and tools for PLC programming are frequently criticised due to their lack of support for integration with other engineering domains, reusability, and reconfigurability, and their inability to effectively validate control behaviour of production systems offline [6]. The management of PLC device configuration is currently completely dissociated from other engineering phases (such as process, mechanical and layout) and from the digital dataset resulting from them. This is mainly due to the lack of capabilities in

- definition of machine control behaviour within Virtual Process Planning tools used to virtually configure and validate automation processes,

- deploying control software on PLC devices directly from the virtual engineering environment and

- the inability to update an engineering dataset when changes occur during commissioning and maintenance phases (e.g., change to PLC control software conducted by accessing the PLC directly, see Section 6.5.5), which creates a critical gap in the CPS systems engineering life cycle that prevents changes and updates carried on the real system to be integrated in the engineering dataset (digital system blue print).

WMG in partnership with FDS is developing an engineering environment (vueOne Virtual Engineering solution) and associated CPS oriented engineering methods that aim at filling this gap by providing automatic control code generation capabilities directly from the vueOne virtual process planning module. The vueOne tools allow definition of control behaviour of actuators, and sensors and sequence of operations at a higher level of abstraction using state-transition diagrams, similar to IEC 61131-3 style SFC (Sequential Function Chart). The sequence checks and interlocks are achieved through the propagation of states between components. The approach allows validation of the control behaviour using 3D CAD based simulation [7].

After virtual validation, the control behaviour is imported to the control code generation module, which facilitate the mapping of resource-specific function blocks to components and allocation of physical I/O addresses. The control code generation module allows generation of both PLC code and associated human machine interface (HMI) screens according to industry standards [8, 9] and configuration for the OPC-UA and REST server for run-time machine monitoring. The generated code is well structured (see Section 11.1.5 for control code architecture) and support functions such as operating modes (i.e., manual, step by step, dry run, automatic and return to initial position), safety interlocks, and fault diagnostic and operator messages. Currently, the code generation module supports Siemens, Schneider, and PLCopen platforms as well as IoT style embedded controllers. Note that the use of the Arrowhead Framework to retrieve OPC-UA client server connection data in order to monitor PLC devices is briefly described in Section 6.5.3.

An additional engineering scenario focusing on PLC device configuration is investigated which focuses on direct deployment of control software to PLC devices using web service and the Arrowhead Framework. Currently the deployment of PLC control software onto the Siemens PLCs used for this project necessitates a direct connection between the PLC and a laptop running the

Siemens STEP 7 environment, which practically means that PLC configurations management scenarios can only be achieved using the vueOne Mapper module. In the future, a DIN mounted server module link to the PLC itself may host the deployment software implemented using Siemen proprietary SDK.

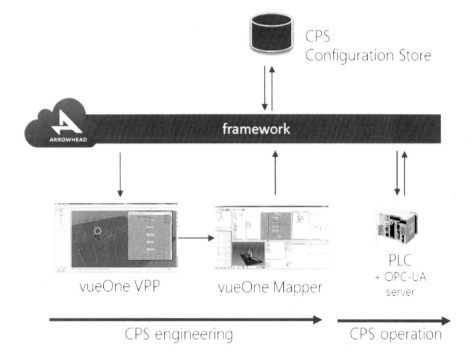

FIGURE 11.4
Deployment and management PLC control configuration.

In an ideal configuration the PLC device (or a server module linked to it) would directly subscribe and pass new control configurations to the Arrowhead Framework ConfigurationStore service when the PLC configuration is changed locally. The PLC device can also retrieve updated configuration if/when available. In practice, however, the ConfigurationStore consumer software is deployed on laptops that control engineers connect to PLCs in order to update or install new control code.

11.1.5 PLC control architecture and deployment scenarios

The research conducted at WMG, University of Warwick, focuses on a data-driven approach [1, 2] to the design and implementation of PLC and distributed control code architecture, which provides a mean to

1. reenforce the concept of component-based engineering methodologies (see Section 6.3) and

2. allow to preserve a vendor independent representation of the control information across the complete CPS system life cycle and

3. facilitate the automatic generation and management of the control code during commissioning and operations phases of the life cycle.

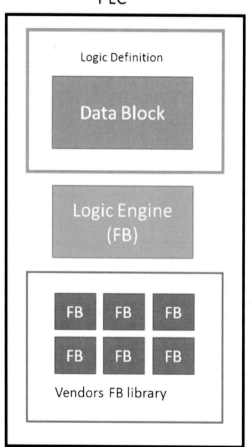

FIGURE 11.5

Data driven control code architecture.

The control architecture designed and implemented by WMG is illustrated in Figure 11.5, which includes a data model or data block (PLC memory array) which contains the control layout of a particular system. The information in the data block includes the definition of components (i.e., state-transition diagrams) as well as the sequence conditions for all components that compose the system, and the interlock conditions to prevent mechanical clash in both manual and automatic modes of operations. The data block is a non-vendor specific representation of a system's control layout that can be used by various engineering tools in order to provide a variety of user/engineering specific views e.g., timing/sequence diagrams, state-transition diagrams (STDs), sequence and interlock chart, XML output, etc.

The logic engine is the executable part of the control code software, which orchestrates system by interpreting the description of the system defined within the data block. The logic engine is configuration independent as its entire source code is generic and remains unchanged for any system configuration. The logic engine is composed of three main modules: operating mode manager, component orchestrator, and fault manager which respectively control the modes of operation and cycle types, check all sequence and interlock conditions and updates the states of the resource-specific and ensure safe operation of machine and handle fault and fault messages.

The resource-specific function blocks (FB) provide interfaces between the control software and the machine I/Os. FBs write directly to the data block upon events (such as change in state of actuators and sensors) which causes the machine state description to be updated and subsequently scanned by the logic engine. The use of resource-specific FBs allows machine builders and end-users to use their own FB libraries, which contributes to code modularity and reuse strategies. The vueOne Mapper module preserves user-specific FBs but implements a wrapper that allows FBs to interact directly with the data block and logic engine.

The motivation for this software architecture is to dissociate the key elements required to achieve the overall device configuration and therefore dissociate the engineering processes that support various aspects of the device configuration (e.g., device firmware/logic engine update, system specific configuration change/update, etc.). This approach also allows generating the control code using standard library components (i.e., FBs), which are driven by the control logic defined in the manufacturing process simulation tools to enable seamless transition from virtual to physical system and feedback of run-time data to the virtual system to facilitate data model calibration and analytics. As the control behaviour is defined in the data model, which can be accessed (i.e. read/write) in runtime, service visualisation and process parameterisation can be attained using human-machine interface devices.

Currently, the deployment of automatically generated control code from the vueOne Mapper module consists in downloading all three components (data block, logic engine, and resource-specific FBs) to the targeted PLC device. However, the data-driven control code architecture potentially provides a basis for additional and more distinctive configuration scenarios.

11.1.5.1   Scenario 1

The logic engine can be updated independently from other control components. This allows scenarios where factory-wide updates of the logic engine for all PLCs devices, can be conducted without impacting on other configuration specific elements of the control (i.e., data block and FBs). The logic engine may be updated in order to improve performances (e.g., scan time) or new functions in the control system (e.g. predictive maintenance, enhanced fault diagnostic, automatic restart, etc.).

11.1.5.2   Scenario 2

If changes to the system's control layout are within a certain scope (i.e., change to safety interlocks and/or sequence conditions only), the data model only or FB only, can be updated. Modification of the data model may result, for instance, from a modification of the control configuration made by control engineers using the vueOne Virtual Process Planning tool. Similarly, FB may be updated regularly (i.e., modification of an FB in order to accommodate a new hardware component) without the need to update the data block or logic engine.

11.1.5.3   Scenario 3

If changes are made to the control layout that require the number or types of states for one or more component to be modified, then both data block and FBs need to be updated and/or deployed to the PLC device. This scenario includes the deployment of new control configuration on an existing PLC (i.e., machine recommissioning or reconfiguration).

11.1.6   Conclusion

In this section, the potential of the Arrowhead Framework to support advanced engineering scenarios has been introduced; the emergence of highly connected, large scale, and complex production systems and processes, will require largely new and advanced engineering capabilities in order to fully realise the potential of CPS systems as defined by the Industrie4.0 vision. The life cycle of CPS systems will blur the boundary between engineering or digital phase i.e., during which a system does not exist physically but is described by an engineering dataset, and the operation phase during which the system exist both physically and digitally through its representation that the data generated by IIoT and connected devices (cyber representation) provides.

The Arrowhead Framework represents a key element in enabling data to be extracted from connected and IIoT devices on the physical system and integrated with digital engineering databases and other source of information. The provision of a framework that readily supports core engineering services (i.e., discovery, orchestration, and security) is essential in encapsulating the complexity associated with complex IT and SOA systems implementation, and therefore in significantly accelerating the implementation of new and advanced engineering solutions. The maintenance scenarios presented above offered the most opportunity for developing Arrowhead oriented solutions, as they allow testing prototype implementations and shadowing maintenance operations without impacting on the production directly. The control code deployment and configurations management are more complex in nature as they impact directly on the production capability. Prototype implementations therefore focused on the integration of various engineering databases and control code deployment and engineering software components rather than on the direct interaction with PLC devices. However, direct access to PLCs can potentially allow for finer grained and potentially automated management of control configurations, which would further close the gap between CPS engineering and operation life cycle phases and therefore offer much more reactive control engineering workflows.

11.2 Manufacturing of electrical cabinets

11.2.1 Introduction

Nowadays, manufacturing plants usually compute production plans using a "no stock" strategy. This is efficient regarding storage and, to some extent, tardiness costs but not at all regarding energy costs. Moreover, with the trend of energy consumption regulation laws and the emergence of the demand-response electricity market, plant managers have to imagine new ways to regulate their energy consumption.

For that purpose, some sensors and a data acquisition chain must be put in place. Measuring energy data allows

(i) the introduction of new KPIs related to energy consumption while evaluating the performance of a plant

(ii) finding machines that consume more energy than the others and change production plans to respect limits or to minimize costs.

In addition, production data can also be obtained from existing systems (like SCADA). Having energy and production data together brings the possibility of computing energy models of production activities and using them to build energy-aware production plans automatically. This way, the "no stock" or "just in time" strategy can be challenged by an energy-aware strategy.

As part of the Arrowhead project, a pilot has been developed aiming at reducing the cost of energy consumption in a Schneider Electric manufacturing plant. This application is based on the same principles as the "Multisource elevator energy optimisation and control" application presented in Section 7.4.1.

The selected site is a Schneider Electric (Sarel) manufacturing plant in Sarre-Union (France), which produces electrical enclosures and electrical cabinets from steel coils and electrical outlets and accessories using injection presses and extrusion machines. The plant is composed of three main activities:

[i] cabinet and cell manufacturing

[ii] plastic electrical appliance injection molding

[iii] plastic tubes extrusion.

The amount of resources spent in the plant is significant: the annual costs of energy and water reach more than 1.6 M€, out of which the major part (62%) is electricity, but the detailed split of this consumption is unknown. The objective was thus to

[i] understand which manufacturing activities consume which amount of energy and

[ii] optimise the production plan and production stock management for global (cost, energy, and time) reduction, taking in account actual dynamic energy prices and the production order book.

The application is a System of Systems, built from the following elements:

- Measurement of energy consumption in the plant. This subsystem consists of an energy measure acquisition chain, based on Schneider Electric wireless sensors and the CEA LINC middleware [10] to manage sensor data.

- Synchronisation of the energy consumption data (one time series per sensor) with production data extracted from a manufacturing execution system (or another legacy system) monitoring the plant. This synchronisation is implemented as a web service in Schneider Electric.

- Characterisation: construction of an energy model of the plant. This is implemented as a web service in Schneider Electric. The current version of this service is focused on plants for which the energy consumption depends mostly on the production activities executed in the plant and not on their interactions or on external factors (such as external temperature, for example). It relies on multivariate regression analysis to identify

  [i] a consumption baseline (i.e., energy consumed in the plant, even when it is idle) and

[ii] the energy required in order to produce a unit of each product reference manufactured in the plant. A more complex "visual" version is currently under study, in order to deal with more complex environments, for which the current regression is insufficient.

- Scheduling future operations based on a description of the plant, the current production order portfolio, the available energy model of the plant, and energy tariff information. The scheduling subsystem receives the description of the plant and the production order portfolio from a manufacturing execution system (or another legacy system); it reuses the energy model built with the previous services; and it obtains the tariff from the Arrowhead-compliant Energy Tariffs shared service. The scheduling subsystem is able to generate different schedules corresponding to different compromises between antagonistic optimisation criteria (timeliness of deliveries, energy costs, storage costs, other production costs). In particular, this enables

  [i] to compare schedules generated with and without considering energy costs and

  [ii] to estimate how much can be gained by considering the energy costs.

- Confirmation and display of the enabled savings. The Arrowhead-compliant Energy Tariffs shared service is used to confirm (and in some cases refine) the energy cost evaluations made by the scheduling subsystem.

Figure 11.6 shows the different software components from synchronisation to the display of the savings as part of the web HMI. The Arrowhead Framework core service: ServiceRegistry is directly consumed by the EnergyTariffWeb service and the web HMI. On the other hand, a manual registration of the Synchronisation and Characterisation web services is performed.

11.2.2 Understanding the consumption: Practical issues

In practice, understanding the consumption law of a plant with respect to the production portfolio has proven to be a real challenge. Indeed, this requires that

[i] we get precise enough energy consumption data,

[ii] we have historical production data recorded as precisely and accurately as possible and that

[iii] we are able to reconcile these two information sources into a single synchronised information system.

For the first point, we have installed the energy measure collection chain at the Sarel plant. The whole chain works correctly. The system was easy to

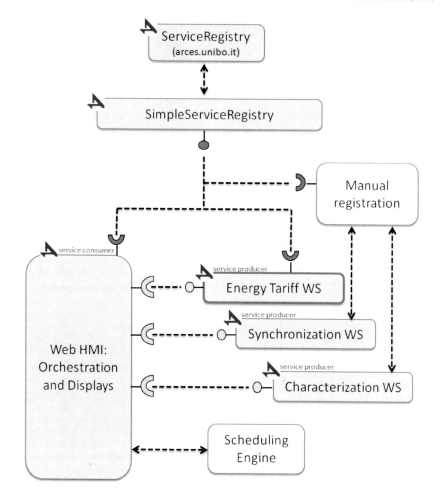

FIGURE 11.6
Manufacturing System of Systems architecture.

install and was quickly up and running. However, the wireless sensors we use provide data only once a given quantity of energy has been measured (this is necessary to enable these sensors to be energetically autonomous). As a result, we cannot have regular measurements, hence some imprecision.

The second point is more difficult, as we are entering the real manufacturing world and the associated information system. In the Sarel plant, manually entered information includes inaccuracies and bias. Start and end times of activities cannot be estimated accurately enough. As a result, the energy consumption analysis must be performed on the basis of aggregated values such as the overall consumption over a 5-minutes period.

The third point concerns finding the right common timescale which is the result of the combination of undersampling and oversampling of the different time series collected for the energy consumption data and the production data. The tuning of the synchronisation is sometimes tricky, and one can easily see that the result could be very sensitive to the precision and accuracy of the construction of historical data. In order to make progress, we have sketched several synchronisation scenarios and used a set of 34 benchmark instances to test different versions of the corresponding web service in the ideal case (with perfect energy consumption and manufacturing production data) and against perturbations of these data. More precisely, we considered the following types of perturbations of the data:

[i] a significant energy consumption baseline (i.e., energy consumed when no manufacturing activity is ongoing in the plant)

[ii] random fluctuations of the energy consumption and/or random errors from the sensors

[iii] reduced frequency of sensor measurements

[iv] imprecise manufacturing execution system data (start and end times of production activities).

Roughly, the obtained results show an appropriate robustness to perturbations (i) and (ii), provided enough measurements are available. Perturbations of type (iii) are well handled, as long as the frequency of sensor measurements remains below the characteristic frequency of the manufacturing process. Perturbations of type (iv) are the most annoying, as the information available for making the analysis quickly degrades with the imprecision of activity start and end times.

11.2.3   Results

We used the above-mentioned benchmark instances to test and compare the efficiency of multiple scheduling techniques (including combinations), as explained in [11]. The best combination, relying on constraint programming, mixed-integer linear programming, and large-neighbourhood search, provides reasonably good solutions in the time limits defined by the benchmark. It remains, however, very sensitive to the number of energy-related variables at each large-neighbourhood search iteration. As a result, it can be expected that improvements will be needed to solve the Sarel case in reasonable computational time. This could take the form of a simple time-based decomposition or of a more complex aggregate optimisation (lot-sizing) module.

When energy costs are not taken into account, our best algorithm provides solutions with a cost that is on average 7% above the best cost reported in the literature. If we except the four biggest instances, the consideration of energy costs improves the use of energy without increasing the tardiness costs. Globally, the total cost is improved. However, the savings are not really important. On these tight benchmark instances from the literature for which avoiding tardiness is already a challenge, only 1% of the energy cost could be saved on average, up to 6% in some instances. This led us to investigate the sensitivity of the results to the energy tariff on the one hand, and on the tightness of production order due-dates on the other hand. Regarding the energy tariff, we started from the current tariffs T1* (typical day/night tariffs as they exist today) and made the following transformations:

- T2*: Reduce the night tariff of T1* by 30% and increase the day tariff by the amount needed so that the total cost of a constant load over 24 hours is unchanged. This enables us to evaluate the capacity to schedule the most energy-costly operations at night. Needless to say, the satisfaction of the production order due dates remains one of the most important optimisation criteria, hence a gain in energy cost will not be accepted if it results in a significant degradation in terms of order tardiness.

- T3*: Reduce the night tariff of T1* by 60% and increase the day tariff by the amount needed so that the total cost of a constant load over 24 hours is unchanged.

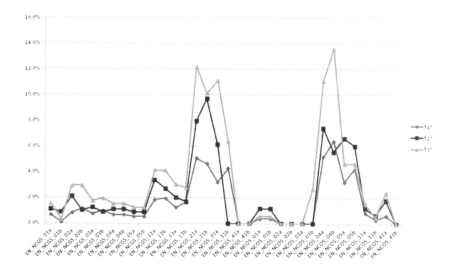

FIGURE 11.7
Sensitivity to energy tariff.

All the different tariffs (five versions of T1*, T2*, T3*) were provided to the tariff server and obtained by the scheduling engine using the Arrowhead Framework EnergyTariffs service. Figure 11.7 shows, for each instance in the benchmark, the gain enabled by energy optimisation under the T1*, T2*, and T3* tariffs. In most cases, the gain increases with the difference between day and night prices, showing the capacity to schedule the most energy-costly operations at night. In some cases, no change is observed and the potential energy gain remains 0. These cases typically correspond to fully loaded factories with either little differentiation in the energy required to produce different references and/or processes and production order due dates which leave little scheduling flexibility. In some cases, the gain goes up or down randomly when we go from T1* to T2* to T3*, reflecting the fact that the algorithm is not able to find the global optimum in the allotted computation time, and that the local optimum that is returned might be of more or less good quality. In some cases, the gain in energy cost exceeds 10% when T3* is used, suggesting that the evolution of tariffs will have a significant impact on the return of investment of the overall solution. Indeed, if we consider a plant with a 1M€ electricity bill per year, 10% means 100k€ per year, which makes the project relevant, while a gain of 2% or 3% would not justify the investment.

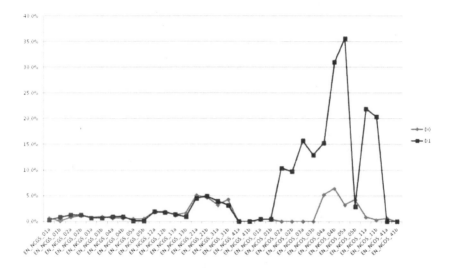

FIGURE 11.8
Sensitivity to production order due dates.

The other element we explored is the sensitivity to the tightness of production order due dates. As already explained, a characteristic of these benchmarks from the literature is that often the due dates are difficult or impossible to satisfy. A consequence of this is that there is little flexibility for energy cost optimisation. Using tariffs T1*, a simple transformation we did to each instance data consists in adding 24 hours to each production order due date. Figure 11.8 shows the results, comparing the energy cost enabled by optimisation with the initial due dates (scenario D0) and with the updated due dates (scenario D1). Its clearly seen that for some instances, the tightness of the due dates in scenario D0 limit the possible energy improvements, while under D1 the gains sometimes become very significant.

Current work is concentrating on the collection and cleansing of actual production data and the full application of the approach to the Sarel use case.

11.3   Asset localisation in mines

Localisation in mining is an important aspect in terms of safety and productivity. By knowing the position of workers and mobile machinery and tools, large gains can be achieved in terms of reduced time needed to locate and retrieve the tools and machines needed for operation. In the case of hazardous events such as fire or tunnel collapse, it is vital to know the position of everyone inside the mine.

Wi-Fi can be used for coarse grained localisation by the use of either signal strength measurements, or just by looking at which access point a device is currently connected to. This gives a rough estimate, usually in the range of 50–70 meters in accuracy. For many cases, this is sufficiently good. Knowing the position of a worker or other entity within 70 meters is far better than having no information about the whereabouts.

However, there are many situations where a much higher accuracy is needed, for example, navigation of autonomous vehicles, rescue operations, logistics optimisation, and others.

Ultra-wide-band (UWB) has been a hot topic in wireless networks and real-time localisation systems (RTLS) for some time now [12, 13]. UWB offers some very interesting properties and capabilities such as high bandwidth, low sensitivity for multi-path interference and ranging. The use of UWB in mining applications has been studied by, for example, Daixian [14].

Autonomous vehicles or smart assets requiring localisation within mines each hosts an Arrowhead Framework local cloud. Operating the autonomous local cloud in this way allows the asset to run as many local services as required, without connectivity to the Internet and head office. When connectivity has been reestablished, the local cloud is accessible through global service discovery. Head office systems or other authorised stakeholders such as maintenance or accounting systems are able to query for relevant information above and beyond current location.

The local cloud on each asset runs a localisation System of Systems which will perform distancing and mine localisation functions. Each UWB node placed on the mine walls will provide a location service. This service will provide the (X, Y) location of the node within the mine, according to a shared map. These nodes do not belong to a single local cloud, but rather, register with the mobile local clouds as the asset moves within the mine. A monitor and control system consumes a trilateration service which calculates the asset's location within the mine. The calculation is based on the position of the static nodes and the distance to each node. This information is collected from the position and the distance services. The diagram in Figure 11.9 shows this System of Systems relationship.

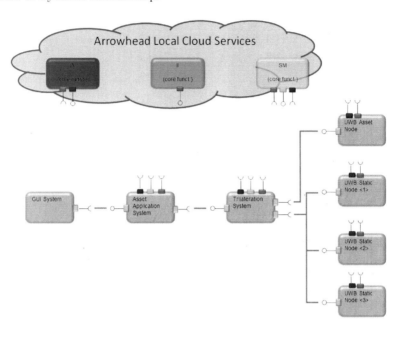

FIGURE 11.9
System of Systems design for Arrowhead mine localisation.

11.3.1 Ultra-wide-band technology

The Decawave DW1000 chip is an IEEE802.15.4-2011 compliant UWB radio [15, 16]. It can perform ranging with an accuracy of $+/-$ 10 cm up to 300 m away with Line-of-Sight (LOS). It has been used along with the Mulle platform to create a ranging node with 6LoWPAN networking capability.

The DW1000 chip supports two forms of distance calculation. It supports using time difference on arrival (TDOA) and time of flight (TOF) distance calculation. It was decided to use TOF between two UWB nodes in order to allow a minimum of two nodes to perform distance ranging.

Because the absolute times of transmission and reception of each node in the pair are used only relative to the same clock domain, the TOF measurements do not need clock synchronisation between two pairs.

The nodes are set up in two modes, as anchors and as tags. There are a total of six messages used in a ranging exchange. Tags are sleepy nodes which broadcast "blink" messages, searching for any anchors within range. This is the discovery message. Anchor nodes are actively listening waiting for "blink" messages from any tag. Once a blink is received tag will begin the ranging handshake by sending an "init" message. The tag responds with a "response" message. The anchor then sends a "poll" message which is responded to with a "final" message. The final message is the last message in the ranging hand-shake. The sixth message is an optional "report" message sent from the anchor to the tag. This message holds the TOF calculated from the four timed mes-sages. This interaction can be seen in Figure 11.10 [16].

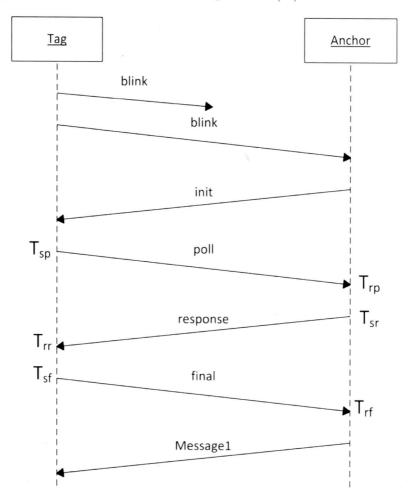

FIGURE 11.10
TOF ranging sequence diagram.

Messages given to the DW1000 for transmission can be provided with a delayed send timestamp. This is critical for reducing the number of messages required for localisation. The "final" message has within its payload the time at which the message is scheduled to be sent. This time is then used in the distance calculation. The formulas 11.1, 11.2, and 11.3 are provided by Decawave. Using the resultant TOF value, the distance is calculated.

$$T_{tof1} = \frac{(T_{rr} - T_{sp}) - (T_{sf} - T_{rp})}{2} \qquad (11.1)$$

$$T_{tof2} = \frac{(T_{rr} - T_{sp}) - (T_{sf} - T_{rp})}{2} \qquad (11.2)$$

$$TOF = \frac{T_{tof1} + T_{tof2}}{2} \qquad (11.3)$$

It is intended that the tag nodes are battery powered and mobile, moving within areas with stationary anchors, performing ranging and providing location information. The anchors, which must always have the radio switched on, listening for "blink" messages, require much more power compared with the tags. In the mine localisation case, the stationary node must be battery powered. This means that they cannot use an always on patter, and must be sleepy. So in this case, the orientation is changed, so that the tags are stationary and battery powered, while the anchors are mobile nodes placed on the mining equipment.

Bibliography

[1] R. Harrison, D. Vera, and B. Ahmad, "Engineering methods and tools for cyber-physical automation systems," *Proceedings of the IEEE*, vol. 104, pp. 973–985, 2016.

[2] D. Zühlke and L. Ollinger, "Agile automation systems based on cyber-physical systems and service-oriented architectures," *Adv. Autom. Robot*, vol. 122, pp. 567–574, 2012.

[3] P. Leitao, J. Barbosa, M.-E. C. Papadopoulou, and I. S. Venieris, "Standardization in cyber-physical systems: The arum case," in *Proc. IEEE International Conference Industrial Technology 2015*, 2015, pp. 2988–2993.

[4] E. Westkämper and L. Jendoubi, "Smart factories-manufacturing environments and systems of the future," in *Proc. 36th CIRP Int. Seminar Manuf. Syst.*, 2003, pp. 13–16.

[5] W. Mahnke, S.-H. Leitner, and M. Damm, *OPC Unified Architecture*. Springer, 2009.

[6] P. Hoffmann, R. Schumann, T. M. Maksoud, and G. C. Premier, "Virtual commissioning of manufacturing systems: A review and new approaches for simplification," in *Proc. ECMS 2010*, 2010, pp. 175–181.

[7] M. Prösser, P. Moore, X. Chen, C.-B. Wong, and U. Schmidt, "A new approach towards systems integration within the mechatronic engineering design process of manufacturing systems," *Int. J. Comput. Integr. Manuf.*, vol. 26, pp. 806–815, 2013.

[8] X. Kong, B. Ahmad, R. Harrison, Y. Park, and L. J. Lee, "Direct deployment of component-based automation systems," in *EEE 17th Conference on Emerging Technologies & Factory Automation (ETFA), 2012*, 2012, pp. 1–4.

[9] "Fast plc structure manual," Ford Motor Company, Tech. Rep., 2013.

[10] M. Louvel and F. Pacull, "Linc: A compact yet powerful coordination environment," in *Coordination Models and Languages: 16th IFIP WG 6.1 International Conference.* Springer, 2014, pp. 83–98. [Online]. Available: http://dx.doi.org/10.1007/978-3-662-43376-8_6

[11] G. German, C. Desdouits, and C. Le Pape, "Energy optimization in a manufacturing plant," in *Proc. ROADEF Annual Conference.* ROADEF, 2015.

[12] H. Liu, H. Darabi, P. Banerjee, and J. Liu, "Survey of wireless indoor positioning techniques and systems," *IEEE Transactions on Systems, Man, and Cybernetics, Part C (Applications and Reviews)*, vol. 37, no. 6, pp. 1067–1080, Nov 2007.

[13] G. Deak, K. Curran, and J. Condell, "A survey of active and passive indoor localisation systems," *Computer Communications*, vol. 35, no. 16, pp. 1939–1954, 2012. [Online]. Available: http://www.sciencedirect.com/science/article/pii/S014036641200196X

[14] Z. Daixian and Y. Kechu, "Particle filter localization in underground mines using uwb ranging," in *Proc. International Conference on Intelligent Computation Technology and Automation (ICICTA), 2011*, vol. 2, March 2011, pp. 645–648.

[15] Decawave, "Analog transmitter DW1000 high level block diagram - DW1000 IEEE802.15.4-2011 UWB transiver," http://www.decawave.com/sites/default/files/resources/dw1000-datasheet-v2.09.pdf, Datasheet.

[16] "Open connectivity foundation." [Online]. Available: http://www.decawave.com/

Index